Nuclear Energy Synergetics

AN INTRODUCTION TO CONCEPTUAL MODELS OF INTEGRATED NUCLEAR ENERGY SYSTEMS

Nuclear Energy Synergetics

AN INTRODUCTION TO CONCEPTUAL MODELS OF INTEGRATED NUCLEAR ENERGY SYSTEMS

A. A. Harms
McMaster University
Hamilton, Ontario, Canada

and

M. Heindler
Technical University of Graz
Graz, Austria

SPRINGER SCIENCE+BUSINESS MEDIA, LLC

Library of Congress Cataloging in Publication Data

Main entry under title:

Nuclear energy synergetics.

Bibliography: p.
Includes index.
1. Nuclear engineering. 2. Nuclear physics. I. Harms, A. A. II. Heindler, M., 1943–
TK9153.N835 1982 621.48 82-16175

ISBN 978-1-4615-9268-6 ISBN 978-1-4615-9266-2 (eBook)
DOI 10.1007/ 978-1-4615-9266-2

Originally published by Plenum Press, New York in 1982
Softcover reprint of the hardcover 1st edition 1982

To
U. M. H.
and
G. Ch. H.

PREFACE

This book is intended as an introductory exposition of those nuclear energy systems concepts which are characterized by an integrated utilization of complementary nuclear processes. Basic to such systems is the notion of *synergism*, which herein implies the cooperative interaction of selected nuclear reactions and system components so as to provide overall advantages not possible otherwise.

While the concept of nuclear energy synergism has in recent years become the focus of an increasing number of conferences, scientific papers, and institutional in-house reports, no text which encompasses the major topics of the subject at a conceptual and analytical level has appeared. It is our aim to present a systematic characterization of these emerging nuclear energy concepts suitable for the senior university student of nuclear science and engineering as well for the active professional involved in nuclear energy systems planning and analysis.

In addressing the subject of nuclear energy synergism we have become most conscious of the tension between realism and vision in nuclear energy technology. As developed here, our perception of nuclear energy synergism is firmly rooted in the present and then seeks to proceed toward a heightened degree of compatibility and efficacy based on an enhanced integration of relevant nuclear processes. It is our view that such conceptual considerations must assume a greater role in the emerging techological orientation of nuclear energy systems planning.

In giving shape and form to our perception of nuclear energy synergism, it was necessary to adopt a perspective and methodology well suited to meet our objective or providing an exposition of both the diversity and the potential of synergetic nuclear energy systems concepts. The current embryonic state of awareness of these emerging systems concepts prompted us to give this text a predominantly pedagogical emphasis at a conceptually fundamental and analytical level. For this reason, we have chosen to use a zero-dimensional lumped-parameter formulation as the main methodological tool of exposition. It followed then that elementary physics and systems considerations combined with a parametric characterization could well be used to describe the various system options while providing compact expressions for the assessment of system performance.

From a broader point of view, we must hasten to add an important qualifier: this book is not to be viewed as a reactor physics text or design manual of synergetic nuclear energy systems. Indeed, we have deliberately avoided the incorporation of details on fission reactor physics, plasma physics, isotope accounting, nuclear data, and station design and operation. This is done for two paramount reasons: first, such details would unavoidably detract from the conceptual flow of synergetic ideas, and, second, either insufficient research exists on some pertinent topics or the appropriate information is already widely available. A Bibliography is included for further clarification of our exposition and for proceeding beyond our chosen level of emphasis.

The organization of this book is in four parts. Chapters 1, 2, and 3 are preparatory in the sense that they place the subject in perspective and cast selected reactor physics and nuclear engineering topics into a form specifically suitable for our purposes. Chapters 4 and 5 contain an analysis of the integration of spallation accelerators and fission reactors. The next part, Chapters 6, 7, and 8, deal extensively with deuterium–tritium based fusion–fission symbionts and hybrids. Finally, Chapters 9 and 10 contain a discussion and analysis of advanced systems options and concepts. Most of Chapter 5 and Chapter 7 emphasize fuel dynamics at a more specialized level and could be deferred in an initial study of this text. Problems suitable as assignments are included at the ends of the chapters. Several appendixes deal with specialized topics and system characterization.

A comment about notation seems appropriate. The analysis of a subject which covers a broad range of related scientific–engineering themes invariably brings with it the problem of notation. We have judged it useful to introduce some revised as well as some new notational forms which emphasize the unifying features of the subject. Self-evident notational simplifications are subsequently made when the context makes its use quite clear. In each instance, however, these changes are noted when first introduced.

In the preparation and writing of this monograph we have repeatedly been made conscious of the indirect role of others. We wish to acknowledge our intellectual debt to W. B. Lewis (Canada), J. D. Lawson (United Kingdom), and other pioneering scientists for early sowing the seed—as it were—for synergetic nuclear energy ideas. Our approach to the subject has been influenced by the writings of L. M. Lidsky (Massachusetts Institute of Technology) and G. M. Miley (University of Illinois). To the scientists at a number of national laboratories, particularly at Brookhaven National Laboratory, Chalk River Nuclear Laboratories, Kernforschung-sanlage Juelich, Kurchatov Institute of Atomic Energy, and Lawrence Livermore Laboratory, we acknowledge their efforts to transform conceptual systems ideas into technically attainable forms. A stay by the senior author at the International Institute for Applied Systems Analysis, and his periodic association with the International Atomic Energy Agency has further contributed both to a broadened perspective of the subject and to a heightened appreciation for the need to conceive and examine advanced energy systems.

It is our pleasure to acknowledge those with whom we have had the pleasure to be in contact as our ideas on nuclear synergism crystallized. This includes a number of very stimulating associates and students, among whom we acknowledge especially C. W. Gordon, W. J. Hartmann, H. W. Lang, K. Niederl, and J. K. Presley. Additionally, we appreciate the detailed review of this work by our associates Dr. E. M. Krenciglowa and Dr. K. F. Schoepf. To the above, as well as to the secretarial staff of the Department of Engineering Physics, the staff of the Engineering Word Processing Centre and to the McMaster University Audio-Visual staff, we express our sincere appreciation.

July 1982

A. A. Harms
McMaster University, Canada
M. Heindler
Technical University of Graz, Austria

CONTENTS

Part III
(D–T) FUSION–FISSION INTEGRATION

Part IV
THE BROADER PERSPECTIVE

Nuclear Energy Synergetics

AN INTRODUCTION TO CONCEPTUAL MODELS OF INTEGRATED NUCLEAR ENERGY SYSTEMS

Part I

AN EMERGING SYNTHESIS

Chapter One

INTRODUCTION

Nuclear energy is often perceived as consisting of two distinct and mutually exclusive technologies: fission or fusion. We develop here the synergetic extension which is based on the integration of these and other complementary nuclear processes in a manner so as to attain desirable characteristics not possible otherwise. The traditional fission or fusion classification of nuclear energy technologies is here taken as the essential starting point for our synergetic approach to the use of nuclear energy.

1.1. NUCLEAR SYSTEMATICS

Of the several hundred naturally occurring nuclei there exists only one, ^{235}U, which will fission under relatively easily attainable conditions. This fission reaction may be represented by

$$^{235}\mathrm{U} + n \rightarrow \nu_{\mathrm{Fi}} n + \sum_i \mathrm{Rp}_i, \qquad U_r \simeq 200\ \mathrm{MeV} \tag{1.1}$$

where n represents a neutron, ν_{Fi} is the average number of neutrons emitted, and Rp_i refers to a reaction product nucleus. The symbol U_r represents the average nuclear energy released and appears in the form of kinetic energy of the reaction products and as electromagnetic radiation.

There are two features associated with Eq. (1.1) which are pertinent to our synergetic theme: this fission reaction is relatively easy to initiate and to sustain and, furthermore, has one of the largest intrinsic nuclear energy gains known. The technology for the production and utilization of its associated energy is well developed and commercially available. However, from the point of view of fuel supply and fuel ash management, there exist two shortcomings: (i) the scarce terrestrial supply of ^{235}U—existing to the extent of 0.72% in natural uranium—is inadequate to sustain, by itself, a substantial fission energy base, and (ii) the management of the radioactive wastes has not been resolved to a generally satisfactory degree.

The recognition that the scarcity of ^{235}U would pose a severe limitation on the expansion of nuclear energy provided the motivation and justification for an early interest in neutronically more efficient fission converter reactors and fission breeder reactors. The reason for an emphasis on neutron economy is that a greater fraction of the fission neutrons would then be available for fissile fuel breeding. This nuclear fuel breeding process is initiated by neutron capture in the more plentiful fertile nuclei ^{232}Th and ^{238}U and may be represented as follows:

$$^{232}\mathrm{Th} + n \rightarrow {}^{233}\mathrm{Th} \xrightarrow[22.3\ \mathrm{m}]{\beta^-} {}^{233}\mathrm{Pa} \xrightarrow[27.0\ \mathrm{d}]{\beta^-} {}^{233}\mathrm{U} \tag{1.2}$$

and

$$^{238}\mathrm{U} + n \rightarrow {}^{239}\mathrm{U} \xrightarrow[23.5\ \mathrm{m}]{\beta^-} {}^{239}\mathrm{Np} \xrightarrow[2.35\ \mathrm{d}]{\beta^-} {}^{239}\mathrm{Pu} \tag{1.3}$$

Here β^- represents beta-decay processes which occur according to the half-lives indicated (m is months and d is days). Eventually, therefore, thermally fissile ^{233}U and ^{239}Pu can replace the diminishing natural ^{235}U as a nuclear fuel. Attendant changes in the actual design of reactors would need to be introduced because there are important and pertinent nuclear differences among these nuclei.

The distinction between a converter reactor and a breeder reactor is determined by the extent to which fissile fuel breeding relative to its consumption takes place in the reactor. A converter reactor is able to replace, by neutron-induced fertile-to-fissile transmutation [Eqs. (1.2) and (1.3)], only part of the fissile fuel concurrently consumed, whereas the breeder reactor eventually generates an excess of fissile nuclei. A reactor is called a burner when fissile fuel breeding occurs to a negligible extent only.

The nuclear fusion energy alternative will initially involve the fusion of a deuterium nucleus, D = ^2H, with a tritium nucleus, T = ^3H, according to

$$\mathrm{D} + \mathrm{T} \rightarrow n + \alpha, \qquad U_r = 17.6\,\mathrm{MeV} \tag{1.4}$$

Here α represents the helium nucleus ^{4}He. Unlike fission, which is initiated by a neutral particle, fusion involves charged particles as reactants requiring, therefore, other forces to initiate and sustain the reaction.

There is an important and frequently overlooked similarity between the basic fission reaction, Eq. (1.1), and the above fusion reaction. While there is sufficient deuterium in the world's oceans to sustain Eq. (1.4), the isotope tritium, being radioactive with a half-life of 12.3 years, does not exist naturally in sufficient quantities to sustain a D–T fusion power program; tritium will therefore need to be bred by nuclear transmutation. This fusile fuel breeding process may involve the fusion neutron of Eq. (1.4) in a capture reaction with naturally occurring though scarce isotopes of lithium:

$$^6\mathrm{Li} + n \rightarrow \mathrm{T} + \alpha, \qquad U_r = 4.8\,\mathrm{MeV} \tag{1.5}$$

and

$$^7\mathrm{Li} + n \rightarrow \mathrm{T} + \alpha + n', \qquad U_r = -2.5\,\mathrm{MeV} \tag{1.6}$$

The former process, Eq. (1.5), proceeds with increasingly higher probability with lower neutron energies, whereas the latter, as indicated by the negative reaction energy U_r, is a threshold reaction.

The initial start-up tritium required for a D–T fusion reactor could be produced in fission reactors by neutron capture in the hydrogen of the coolant or moderator; the continuing tritium supply is expected to be bred in a lithium blanket surrounding the fusion core while the fusion reactor is in operation.

It is expected that following the successful demonstration of D–T-based fusion energy, it will be possible to introduce deuterium-fueled fusion reactors; although this process is more difficult to achieve, it has the advantage of eliminating the need to breed tritium. The D–D fusion reaction proceeds via two essentially equally likely channels:

$$\mathrm{D} + \mathrm{D} \rightarrow {}^3\mathrm{He} + n, \qquad U_r = 3.2\,\mathrm{MeV} \tag{1.7}$$

and

$$D + D \rightarrow T + p, \qquad U_r = 4.1\ \text{MeV} \tag{1.8}$$

where p represents a proton; this pure deuterium-burn cycle may be represented by one equation written as

$$4D \rightarrow {}^3He + T + p + n, \qquad U_r = 7.3\ \text{MeV} \tag{1.9}$$

where U_r is the total nuclear energy associated with this particular fusion cycle.

Additional exoergic fusion reactions exist and will be discussed in subsequent sections.

While the technology for the commercial utilization of fission energy is already in place, considerable research and development continues to be necessary before fusion energy can be realized. The potential depletion of lithium could possibly restrict D–T fusion reactor deployment just as the natural scarcity of ^{235}U limits the large-scale use of nonbreeder fission reactors. Indeed, the analogy that D–D fusion would bypass the lithium exhaustion limitation much as the fast breeder could eliminate ^{235}U scarcity appears evident. Additionally, while neither D–T nor D–D fusion produces radioactive reaction products, it is recognized that the tritium inventory and tritium processing, as well as the neutron-induced radioactivity in the materials present in a fusion reactor blanket, will contribute to a significant radiological hazard.

1.2. REACTION LINKAGES

Though the preceding discussion has identified the dominant nuclear processes relevant to fission, fusion, and breeding, the traditional forms in which the equations are displayed do not focus sufficiently upon the role of nuclear catalysis and reaction linkages which are of vital importance to a synergetic conceptualization. We consider therefore the following more general form of pertinent reactions and allude to the broader implications.

A simple nuclear reaction involving fuel nuclei F_1 and F_2 may generally be written as

$$F_1 + F_2 \rightarrow \sum_i Rp_i \tag{1.10}$$

where Rp_i denotes one of several distinct or identical nuclear reaction products.

The important fission chain reaction in which one of the ν_{Fi} fission neutrons n together with a thermally fissile nucleus Fi is used to sustain a chain reaction is given by

$$\mathrm{Fi} + n \rightarrow n + (\nu_{\mathrm{Fi}} - 1)n + \sum_i \mathrm{Rp}_i \tag{1.11}$$

where the neutron acts like one of the fuel nuclei in Eq. (1.10).

A breeding cycle in which a reaction product neutron is used to breed a fuel nucleus may be represented by

$$\begin{aligned} &F_1 + F_2 \rightarrow \nu_{\mathrm{Fi}} n + \sum_i \mathrm{Rp}_i \\ &\mathrm{Fb} + n \rightarrow F_1 + \sum_i \mathrm{Rp}_i \end{aligned} \tag{1.12}$$

where Fb is a fertile nucleus suitable for fissile fuel breeding and ν is a number equal to or greater than unity. The tritium-breeding D–T fusion cycle is an example of such a reaction system which specifically involves the neutron as a chain carrier; the combined use of the fission-breeding equations, Eq. (1.1) with Eq. (1.2) or Eq. (1.3), is also a special case of this equation.

Combining the fission chain reaction with a breeding process in a fission reactor context yields

$$\begin{aligned} &\mathrm{Fi} + n \rightarrow n + (\nu_{\mathrm{Fi}} - 1)n + \sum_i \mathrm{Rp}_i \\ &\qquad\qquad\qquad\qquad \searrow \text{neutron losses} \\ &\mathrm{Fb} + n \rightarrow \mathrm{Fi} \end{aligned} \tag{1.13}$$

Here we emphasize the important role of neutron conservation in maintaining a closed reaction cycle. If ν_{Fi} is relatively small and/or if neutron losses are sufficiently large, then fewer fissile fuel nuclei Fi are bred than consumed; consequently, the balance of fissile fuel has to be supplied externally if the chain reaction is to be sustained. Indeed, whether or not the entire cycle of Eq. (1.13) can—in a time-integrated sense—be closed represents the distinction between a breeder and converter fission reactor.

This last point alludes to a fundamental aspect of fission and D–T fusion energy: neutrons are necessary for fuel-breeding purposes. Hence, in order to aid this neutron economy, one may include neutron multiplication processes with nuclei M,

$$n + M \rightarrow xn + \sum_i \mathrm{Rp}_i \tag{1.14}$$

in the reaction scheme requiring, however, that, on balance, $x > 1$.

In addition to the use of neutrons for breeding and multiplication purposes, neutrons are also useful for transmuting certain radioactive reaction products Rp_j into less hazardous, shorter-lived, or even stable form:

$$n + \mathrm{Rp}_j \rightarrow \sum_{i \neq j} \mathrm{Rp}_i \tag{1.15}$$

With the role of neutron as a nuclear catalyst so vital, a question of interest concerns the inclusion of additional reactions which are particularly neutron rich. It is known that a high-energy proton can interact with a heavy nucleus releasing a substantial number of neutrons as a result of direct nuclear cascade and ensuing evaporation processes. The entire reaction process may be compactly written as

$$p + Z \rightarrow \nu_{\mathrm{sp}} n + \sum_i \mathrm{Rp}_i \tag{1.16}$$

where p represents the high-energy accelerator-produced proton interacting with a high-Z nucleus and ν_{sp} denotes the average number of neutrons produced per spallation event. This process is generally called a spallation reaction.

It is apparent then that the various reactions discussed above could be selectively combined in various forms with three dominant objectives:

i. breed fuel,
ii. produce energy,
iii. transmute hazardous wastes.

The degree to which these three objectives can be realized is dependent on the extent to which the associated reactions can be sustained and integrated. As suggested in Figure 1.1, various and diverse reactions are potentially available for examination for this purpose.

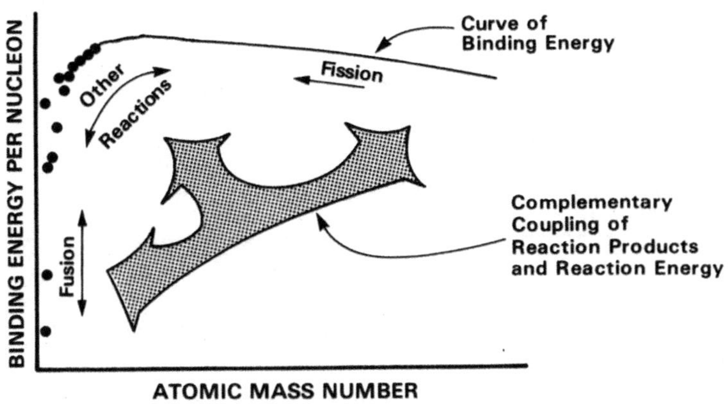

FIGURE 1.1. Schematic representation of the coupling of various nuclear reactions superimposed on the curve of binding energy.

1.3. HISTORICAL DEVELOPMENTS

Though the early developmental period of the civilian nuclear power program was marked by considerable optimism about the eventual positive impact of nuclear energy, there nevertheless existed apprehensions about the adopted developmental program emphasis. These concerns were prompted not only by the evident scarcity of ^{235}U but also by hazards associated with fuel and waste management schemes as well as by the possibility that neither the fast-breeder nor fusion energy could be available at the appropriate time and under acceptable terms. The ready accessibility of fission reactor technology suggested its use to the largest extent possible; the notion that neutrons could play a catalytic role was also recognized early.

The earliest published record which suggested a definite and, in principle, workable integration between a fission reactor and another device—that is, a synergetic nuclear energy system—is attributable to W. B. Lewis (1952)* of the Chalk River Nuclear Laboratories. In an in-house report dated August 25, 1952, Lewis proposed that an accelerator be employed specifically to produce neutrons which would then be used to breed fissile fuel for fission reactors. The accelerator would be supplied by electric power provided by a fission reactor and, in return, provide the companion reactor(s) with bred nuclear fuel. We illustrate this concept in Figure 1.2. The reaction linkages for this accelerator scheme can be readily represented in terms of the preceding notation. Listing only the essential members of the reactions and specializing to the thorium fuel

* Appendix A provides the citation source of references used in this text.

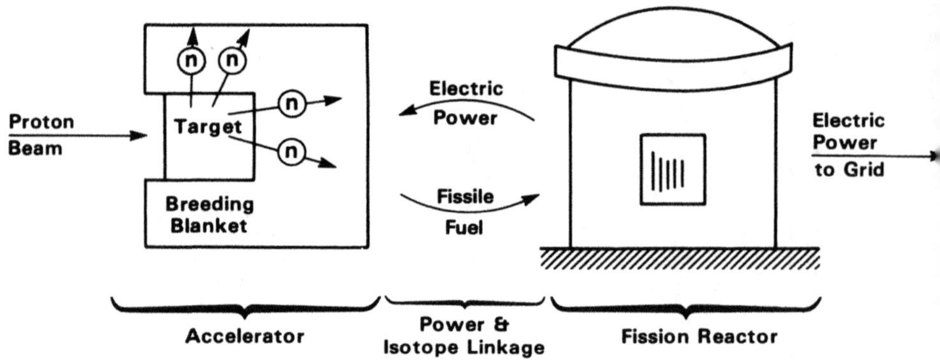

FIGURE 1.2. Illustration of a spallation accelerator coupled to a companion fission reactor as suggested by W. B. Lewis (1952).

cycle, the reaction system may be described by the following reaction linkage representation:

$$
\begin{array}{ll}
{}^{233}\mathrm{U} + n \rightarrow n + (\nu_{\mathrm{Fi}} - 1)n + \sum_i \mathrm{Rp}_i & \left.\right\} \text{Fission reactor} \\
{}^{232}\mathrm{Th} + n \rightarrow (\ ) \rightarrow (\ ) \rightarrow {}^{233}\mathrm{U} & \left.\right\} \text{Accelerator blanket} \\
p + Z \rightarrow \nu_{\mathrm{sp}} n + \sum_i \mathrm{Rp}_i & \left.\right\} \text{Accelerator target}
\end{array}
\quad (1.17)
$$

This scheme suggests an effective linking of accelerator technology with fission technology.

One year later, the California Research Company prepared a proposal for the development of what was called a "driven thermonuclear reactor." The interesting feature of this proposal was to use a preionized deuterium or tritium plasma which would serve as a target for light ions such as deuterons; the ion beam would be produced by an accelerator and the plasma—containing tritium—was to be surrounded by a blanket containing ^{238}U. This ^{238}U was expected to fission by the 14-MeV D–T fusion neutrons and thus act as a multiplier of neutrons. The resultant neutrons would then breed ^{239}Pu and tritium, Figure 1.3. This facility therefore breeds both fissile and fusile fuel and represents an interesting synergism of fusion, fission, and breeding processes. The complementarity of these reactions is obvious, as is evident in the following representation of the

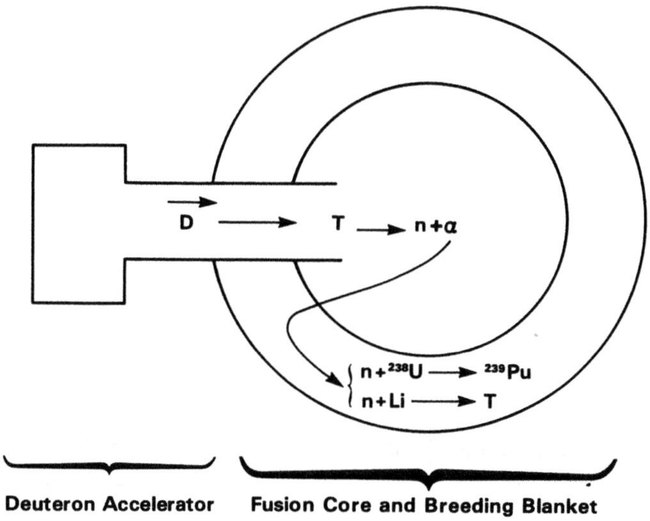

FIGURE 1.3. Depiction of an accelerator-driven fusion reactor and a breeding blanket as proposed by the California Research Company (1953).

possible reaction scheme, again showing only the essential features:

$$
\begin{array}{ll}
D + T \rightarrow n + \alpha & \} \text{ Beam-plasma interaction} \\
{}^{238}U + n \rightarrow \nu_{Fi} n + \sum_i Rp_i & \} \text{ Blanket neutron multiplication} \\
\left\{ \begin{array}{l} n + Li \rightarrow T + \alpha \\ n + {}^{238}U \rightarrow (\ ) \rightarrow (\ ) \rightarrow {}^{239}Pu \end{array} \right. & \} \text{ Blanket breeding} \\
& \} \text{ Fission reaction}
\end{array}
\qquad (1.18)
$$

This system closes the fusion fuel cycle and provide fissile fuel for fission energy purposes. In contrast to the proposal of combining accelerators with fission reactors, this concept represents a merging of accelerator technology with fusion technology.

The next pertinent documented contribution to nuclear energy synergetic ideas appeared in 1955 and is attributable to J. D. Lawson of the Harwell Laboratories. Lawson (whose name is associated with the Lawson criterion in fusion physics) considered a D–T–^{6}Li fusion reactor and introduced what he called the "fusion–fission system." The noteworthy point emphasized is that regardless of how D–T fusion is to be accomplished, the 14-MeV fusion neutrons are to fission the ^{238}U in the surrounding blanket and in so doing, Lawson calculates, the net result will be an

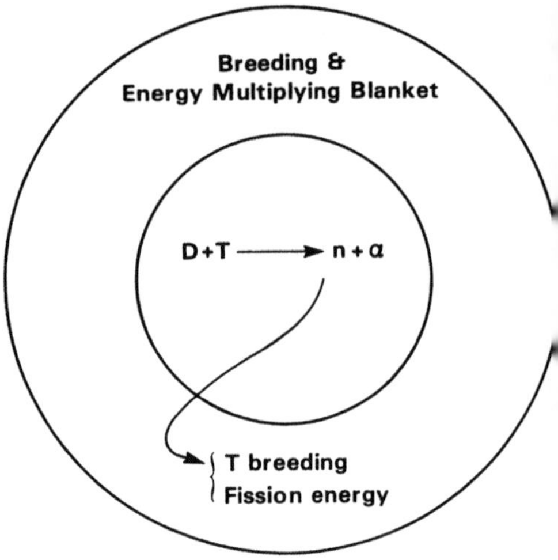

FIGURE 1.4. Schematic illustration of the J. D. Lawson (1955) fusion–fission system.

increase in energy to 122 MeV per D–T fusion event; the concept of nuclear energy multiplication was thus introduced.

A reaction representation of the Lawson scheme, Figure 1.4, is provided by the following reaction schematic:

$$\begin{array}{ll} D + T \rightarrow n + \alpha & \} \text{ Fusion core} \\ \left\{ \begin{array}{l} n + {}^{6}\text{Li} \rightarrow \alpha + T \\ n + {}^{238}\text{U} \rightarrow \nu_{Fi} n + \sum_i \text{Rp}_i \end{array} \right. & \} \text{ Blanket} \end{array} \tag{1.19}$$

As is evident, this provides for a direct linking of fusion technology with fission technology.

The above three examples indicate the variety of early synergetic nuclear energy ideas. While these ideas initially generated little interest—caused in part by the general level of secrecy and the confidence in the main-line reactors—the program of declassification and the broader exchange of scientific ideas combined with perceived difficulties of first-generation fusion and fast breeder reactors have more recently prompted considerable research and development activity on the subject. The contemporary emphasis on nuclear energy synergism began in 1969 when two papers appeared on the subject, one by L. M. Lidsky and one by H. Jung (Appendix A).

The preceding discussion served the purpose of emphasizing some salient features of selected synergetic nuclear energy aspects and also alluded to the potentially large number of distinct synergetic systems concepts which might be conceived. One emerging feature—which has been particularly highlighted by the three historical synergetic examples—is that the technologies of fission, fusion, and accelerators need not be viewed as independent but intrinsically linked; we suggest, therefore, the existence of an underlying nuclear energy systems continuum as illustrated in Figure 1.5. Nuclear energy synergism thus rests fundamentally on the broader and enhanced utilization of selected complementary nuclear processes. One useful image to adopt is that the fundamental basis of nuclear synergism is the judicious rearrangement of nucleons as governed by the curve of binding energy (Figure 1.1).

It seems appropriate, at this juncture, to project the basic synergetic reaction formulations into a context which provides a more specific indication of the options and characteristics of the resulting systems concepts. One may consider the following starting point: With (thermal) fission technology essentially developed, one possible evolutionary approach might be to render this technology more effective and more useful by developing either accelerators or fusion reactors in an integrative support capacity. Based on such a strategy, one synergetic nuclear energy ideal which relates to a more immediate realization of nuclear energy synergism

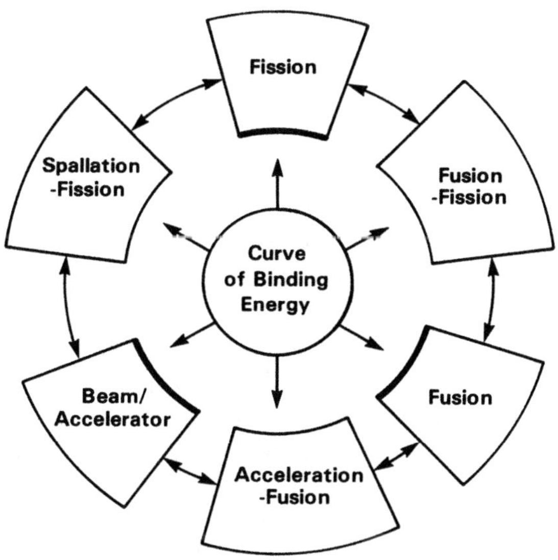

FIGURE 1.5. Graphical characterization of nuclear energy technologies as they appear in retrospect up to 1955.

may be formulated in the following way: the fission reactor, being energy rich but neutron poor, would supply a fraction of its energy—as necessary—to an accelerator or fusion device and in return receive fuel to support its principal objective of energy production. While the energy-for-fuel exchange is central, an additional complementarity associated with such a symbiosis becomes apparent. This has to do with the feature that since both the spallation neutrons and fusion neutrons possess higher energies than the fission neutrons; there exists the physical basis to perform functions not possible in a fission reactor domain. Hence, a question of interest is to determine how these functions can best be realized in a practical configuration.

Basic to the incorporation of a beneficial fusion–fission–spallation symbiosis is the fullest possible exploitation of the blanket which surrounds a fusion core or spallation target. Both experimental and analytical considerations suggest that there are indeed many distinct blanket design options which could serve a variety of purposes. Fundamental to these various purposes are the following characteristics associated with these neutron-driven blankets:

(i) the higher initial neutron source energies allow a further enhancement of neutron multiplication, and
(ii) the higher initial neutron source energies allow more latitude for neutron energy spectrum shaping to enhance selected transmutation functions.

It is desirable that the blankets operate as subcritical assemblies eliminating many of the vital considerations of reactor safety and control.

With the above considerations providing the basis for breeding fissile and/or fusile fuels, in a broader context one can identify a number of specific as well as general systems characteristics.

A fission reactor suitably coupled by energy and isotope linkages to either a spallation accelerator or a fusion reactor may be viewed as a synergetic breeder system. However, in contrast to a conventional fast breeder, the effective breeding capacity of a synergetic breeder can be exceedingly large. Furthermore, operating in a symbiotic relationship with a fission reactor, the fusion facility or the spallation facility by itself need not be a net energy producer. These support systems may be viewed as representing a fractional energy burden on the fission reactor, much as a uranium mine or fuel-enrichment plant represents an energy load on the fission reactor. The issue then is not to allow the energy cost of the externally supplied non-fission-generated neutrons to become too large.

The spallation and fusion blankets offer some particularly promising possibilities of nuclear fuel management. These possibilities are associated with the functions of neutrons entering the blanket. By suitably haping the neutron spectrum it may be possible to selectively transmute uclei in accordance with desired objectives. Among these we cite destruction of fission products, destruction of spallation products, and, of course, reeding of fissile and fusile fuel. This latter function may potentially e done *in situ*, reducing the need for nuclear fuel reprocessing.

The existence of an ample supply of neutrons suggests that many tructural components for reactors and blankets could be designed with ncreased durability, expanded reliability, reduced radioactivity, and nhanced safety—rather than neutron economy—as the dominant design riteria. Furthermore, the full development of the blanket fuel breeding otential would greatly expand the potential of accessible nuclear fuel y allowing the conversion of the accumulated depleted fertile materials rom enrichment plants.

One particularly appealing feature of these synergetic nuclear systems s that they provide the essential basis for the development of syntonic uclear energy parks. These energy parks would be fuel self-sufficient, ossess an intrinsic nuclear waste transmutation capacity, and reduce ertain nuclear hazards.

We will address this and related topics in the following chapters.

PROBLEMS

1.1. List specific examples of Eq. (1.10) and Eq. (1.14).

1.2. Identify on the curve of binding energy typical fission, fusion, and transmutation reactions.

1.3. Describe the role of neutrons in fission energy production, in fissile and fusile fuel breeding and in nuclear waste transmutation.

1.4. Identify the motivation for calling fusion reactions, in particular D–T, "neutron rich," and fission reactions "neutron deficient" although ν_{Fi} is considerably greater than ν_{Fu}.

1.5. What is the interest of integrating a fusion or spallation neutron source into a fission reactor economy? What figures of merit would you introduce to describe those characteristics of fusion and spallation neutron sources which may identify their relative merits in a synergetic system?

1.6. Provide a physical characterization and corresponding mathematical description which emphasize the distinction between neutron–nucleus and ion–ion interactions.

Chapter Two
NUCLEAR PROCESSES

The analysis of synergetic nuclear energy systems will be considerably aided by the recognition of the role of pertinent nuclear processes. In this chapter we identify and describe the dominant nuclear processes of interest to us here, emphasizing the descriptive aspects of the subject as they relate in our nuclear synergetic theme.

2.1. FISSION: NEUTRON YIELD, ENERGY YIELD, NEUTRON SPECTRUM

As suggested in the preceding chapter, the number of unbound neutrons is the dominant consideration to the maintenance of fission chains and neutron-induced transmutation processes. The important neutron-producing reactions are fission, spallation, fusion, and neutron multiplication.

In Chapter 1 we introduced a particular fission reaction involving the naturally occurring ^{235}U nucleus. Other nuclei possessing similar fission characteristics are ^{233}U, ^{239}Pu, and ^{241}Pu. Though these do not exist in nature, ^{233}U and ^{239}Pu may be produced by nuclear processes as previously indicated; ^{241}Pu is produced by successive neutron capture from ^{239}Pu. Further, many other nuclei will fission when the kinetic energy of the incident neutron exceeds a specific threshold; the more pertinent nuclei which undergo such fast fission are ^{232}Th and ^{238}U. The reaction energies U_r for these fissionable nuclei are close to 200 MeV (Table 2.1).

TABLE 2.1
Reaction Energies of Selected Fissionable Nuclei

Isotope	Reaction energy, U_r (MeV)
^{232}Th	184 (fast fission)
^{233}U	190 (thermal fission)
^{235}U	193 (thermal fission)
^{238}U	194 (fast fission)
^{239}Pu	199 (thermal fission)

The fission reaction representations Eq. (1.1) and, more generally, Eq. (1.11) suggest that fission may be viewed as the absorption of a neutron by a fissionable nucleus followed by the emission of an average of ν_{Fi} fission neutrons. This number increases monotonically with the energy of the neutron inducing the fission, and depends upon the type of nucleus undergoing fission. A functional representation generally found useful is given by

$$\nu_{\mathrm{Fi}} \simeq a_1 + a_2 E \tag{2.1}$$

with a_1 in the range of 2–2.5, depending upon the nucleus, and a_2 approximately 0.15 for E in units of MeV.

It is known that not every neutron absorbed in a fissionable nucleus will lead to fission; the neutron may simply be captured by the nucleus transmuting it into another isotope of the same element. These two reaction channels may be represented by

$$\mathrm{Fi} + n \begin{cases} \xrightarrow{\varepsilon_{\mathrm{fi}}} \nu_{\mathrm{Fi}} n + \sum_i \mathrm{Rp}_i \\ \xrightarrow{\varepsilon_c} \mathrm{Rp} + \cdots \end{cases} \tag{2.2}$$

where $\varepsilon_{\mathrm{fi}}$ is the branching ratio for fission occurring given that neutron absorption has taken place and ε_c refers to radiative neutron capture under the same conditions.

Upon neutron absorption, we have $\varepsilon_{\mathrm{fi}} + \varepsilon_c = 1$, and using $\alpha = \varepsilon_c/\varepsilon_{\mathrm{fi}}$ we may write

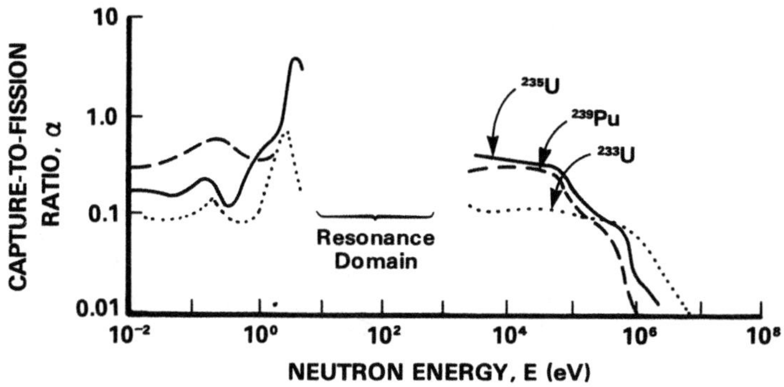

FIGURE 2.1. Capture-to-fission probability ratio as a function of neutron energy.

$$\varepsilon_{\text{fi}} = \frac{1}{1+\alpha}, \qquad \varepsilon_c = 1 - \frac{1}{1+\alpha} \tag{2.3}$$

where α represents the capture-to-fission probability ratio as widely used in reactor physics. This capture-to-fission ratio is known to be strongly dependent on the kinetic energy of the incident neutron as illustrated in Figure 2.1. This figure also suggests that at low and intermediate energies, a significant fraction of neutrons and fissile nuclei are lost by this parasitic capture process.

There are two relevant points to add. The first is that, whereas the reaction energy associated with a fission reaction has been designated as U_r, the total energy yield per neutron absorption in a fission reactor includes other associated processes. We will use the symbol U_{Fi} for this latter purpose. The second point is that the important neutron yield parameter is not so much the average number of neutrons released per fission, ν_{Fi}, but the average number of neutrons emitted per neutron absorbed in a fissile nucleus. This net neutron yield parameter will be designated by η_{Fi} and, as evident from Eq. (2.2), for the case of interest here, is given by

$$\eta_{\text{Fi}} = \nu_{\text{Fi}}\varepsilon_{\text{fi}} = \frac{\nu_{\text{Fi}}}{1+\alpha} \tag{2.4}$$

Note that these functions depend on the energy of the incident neutron.

Though ν_{Fi} is, in general, smoothly and monotonically increasing with energy, η_{Fi} is, by virtue of its dependence on α, a strong function of the energy of the incident neutron as well as of the specific fissionable nucleus; see Figure 2.2. The useful feature is that for neutron energies in the intermediate and higher energy range, the neutron yield increases

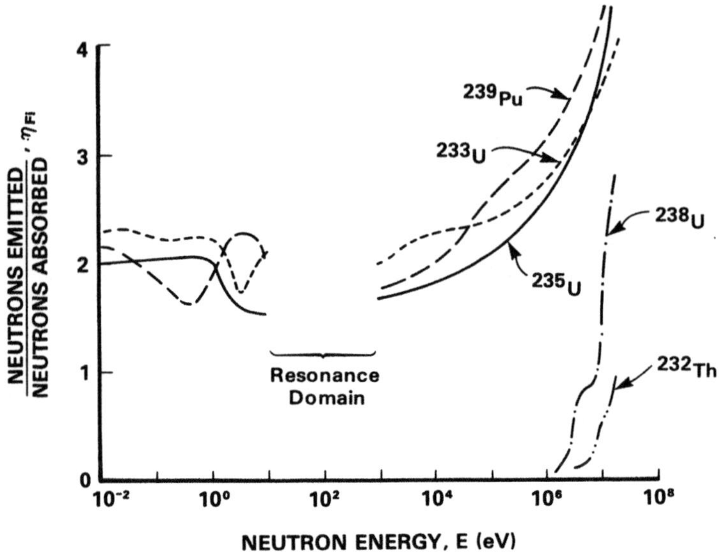

FIGURE 2.2. Fission process neutron yield as a function of incident neutron energy.

considerably with neutron energy; this neutron multiplication capacity of "hot" neutrons will continue to be an important consideration throughout this text.

Another important feature of fission neutrons is their distribution in energy. When neutrons first appear as a result of fission, their distribution

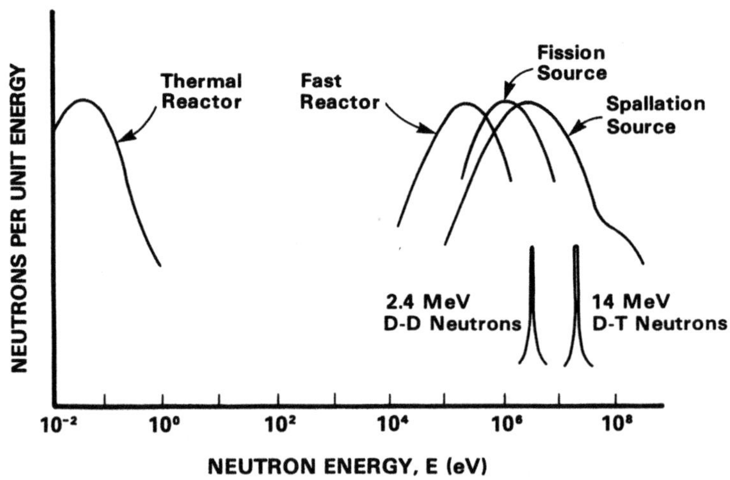

FIGURE 2.3. Schematic depiction of neutron spectra associated with selected reactions and reactors for arbitrary neutron source strength.

follows a characteristic fission source distribution which possesses a peak at ~0.8 MeV and has an average of ~2 MeV. This neutron source energy spectrum changes towards lower energies very quickly as the neutrons lose energy by scattering collisions with the nuclei in the medium. We illustrate in Figure 2.3 the fission source distribution as well as the neutron energy spectrum in a fast and thermal fission reactor; the other distributions shown will be discussed subsequently.

Finally, in addition to the ν_{Fi} neutrons appearing directly and immediately in the fission process, some of the radioactive fission products also lead to the emission of neutrons. Although these delayed neutrons contribute less than 1% to the total neutron population, they are of considerable importance to the control of nuclear reactors.

2.2. SPALLATION: NEUTRON YIELD, ENERGY YIELD, NEUTRON SPECTRUM

A high-energy proton beam incident on a target consisting of medium-to-heavy nuclei is known to lead to the release of a large number of neutrons as well as some protons. This type of reaction is called nuclear spallation and was introduced in the previous chapter. The associated neutron production process may be explained in terms of two distinct phenomena. Initially, the proton knocks out nucleons and nuclear fragments—that is, both neutrons and protons—by direct cascade interaction. Then, following this initial process, the residual excited nucleus ejects additional neutrons by nuclear evaporation. The high-energy cascade nucleons will invariably lead to secondary reactions with the nuclei in the target and thus contribute to further neutron emission. We depict this spallation process in Figure 2.4.

The total neutron yield in a spallation reaction is a function of the incident proton energy as well as of the target nucleus. Additionally, the target size is important because it determines the possible number of secondary neutron multiplication processes. We display in Figure 2.5 the net number of neutrons produced per incident beam proton, η_{sp}. The results displayed in this figure are experimentally confirmed for small targets, typically cylindrical targets of ~20 cm radius and ~60 cm length, while the large target results are based on calculations. An analytical representation of the number of neutrons released per beam proton absorbed in a small nonfissile target may be written as

$$\eta_{sp} \simeq a_3(A + a_4)(E + a_5) \tag{2.5}$$

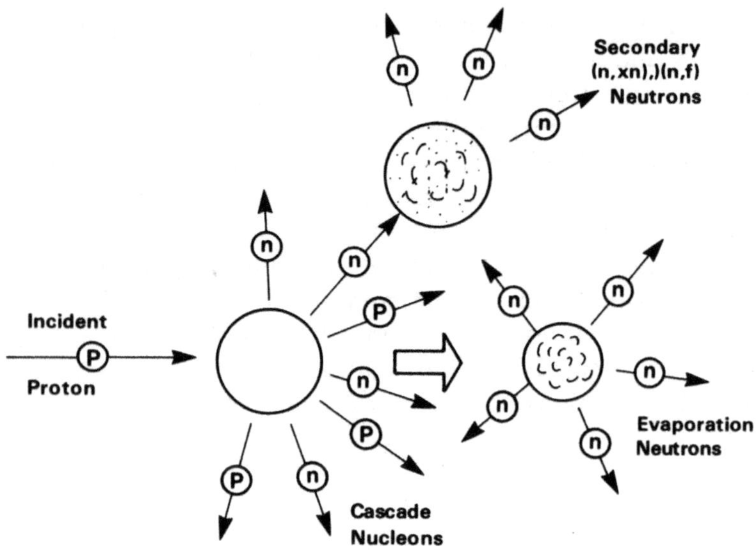

FIGURE 2.4. Illustration of a proton-induced spallation reaction.

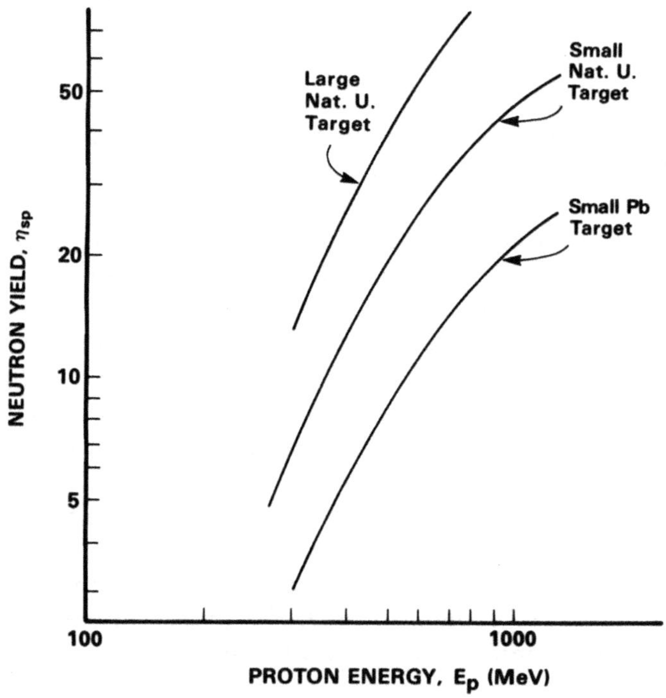

FIGURE 2.5. Approximate neutron yield per incident proton in various spallation targets as a function of incident proton energy.

where the a_i's are empirical parameters.

The energy spectrum of neutrons generated in a spallation accelerator target contains two distinct components. The more dominant part has a peak in the range of 2 to 4 MeV with a shape somewhat harder than a fission spectrum and is attributable to the evaporation process; the cascade neutrons may possess energies up to values close to the energy of the incident proton and form the high-energy hump of the spallation neutron distribution (Figure 2.3). This latter group of neutrons is characterized by considerable anisotropy with a predominant forward peak while the former component can be assumed to be isotropic.

A fraction of the kinetic energy of the incident proton supplies the required dissociation energy for the spallations nucleus. However, the resultant high-energy neutrons will initiate exoergic processes and thus generate thermal energy particularly as a result of fast fission. Since this heat generation will, in a first approximation, vary with the neutron population, the energy yield in the target will similarly vary with incident proton energy, target material, and target size. This energy yield is shown in Figure 2.6, where we also indicate the extent to which the accelerated proton energy is multiplied in the several targets represented by $G_{sp} = U_{sp}/E_p$. The total target energy release per proton, U_{sp}, may again be

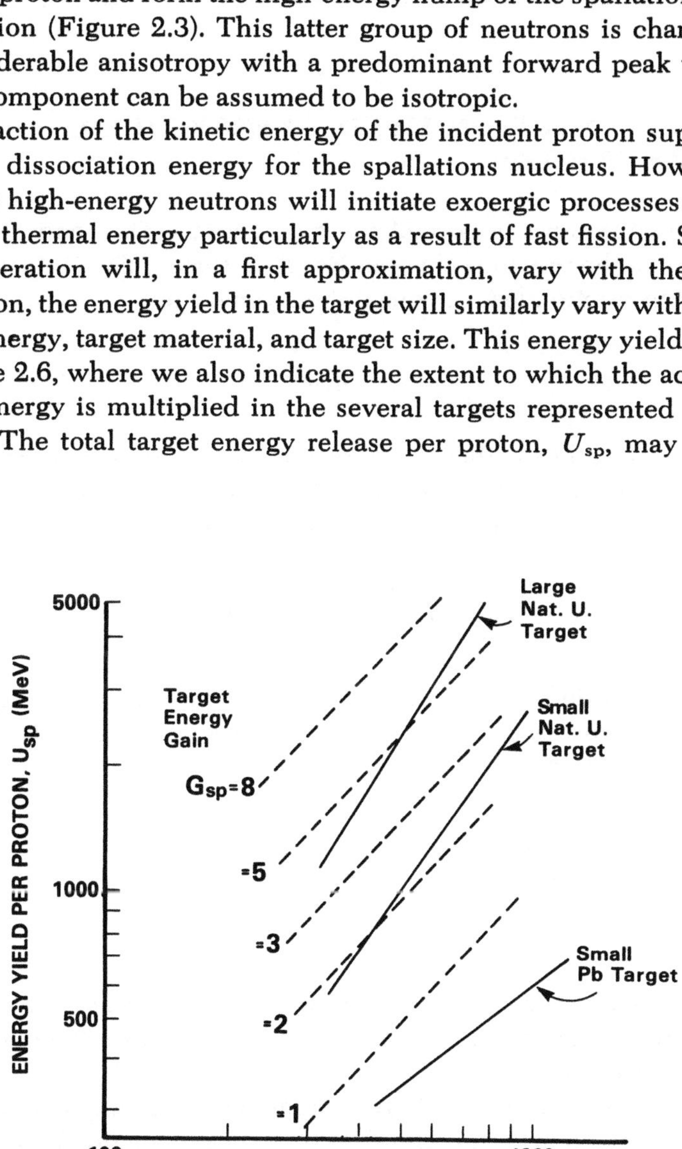

FIGURE 2.6. Energy production in a spallation target as a function of proton energy. The term G_{sp} refers to the multiplication of the proton energy in the target.

represented by an empirical relation of the form

$$U_{sp} \simeq a_6 E^{a_7} \tag{2.6}$$

with parameters a_6 and a_7 extractable from Figure 2.6 and clearly dependent upon the details of the target and the proton energy.

2.3. FUSION: NEUTRON YIELD, ENERGY YIELD, NEUTRON SPECTRA

Two of the most important fusion reactions, D–T and D–D, were introduced in the previous chapter. For D–T fusion, Eq. (1.4) shows that every fusion reaction leads to one unbound neutron:

$$\nu_{Fu} = \eta_{Fu} = 1 \tag{2.7}$$

Here, as in virtually all other fusion reactions, the reaction energy appears as kinetic energy since the first excited level of the reaction product ions is at a sufficiently higher energy; for D–T fusion, we get

$$U_r = E_n + E_\alpha = 17.6\ \text{MeV} \tag{2.8}$$

According to basic kinematic considerations, the reaction energy is split between the two reaction products n and α according to their relative mass ratios as follows:

$$E_n = U_r\left(\frac{m_\alpha}{m_\alpha + m_n}\right) = 14.1\ \text{MeV}, \qquad E_\alpha = U_r\left(\frac{m_n}{m_\alpha + m_n}\right) = 3.5\ \text{MeV} \tag{2.9}$$

Thus, in the case of D–T fusion, 80% of the reaction energy is carried by the neutron while, in contrast, in a fission reaction only ~3% is associated with the ν_{Fi} fission neutrons.

The distribution in the relative energy before interaction between D and T ions leads to a slight spread of the neutron energy distribution as indicated in Figure 2.3.

The neutron yield associated with various D–D burn fusion modes can be illustrated as follows. Three distinct and idealized fusion cases may be identified:

(a) PURE-D fusion: (D–D fusion occurs according to two essentially equally likely channels):

$$\begin{aligned} D + D &\rightarrow {}^{3}He + n \\ D + D &\rightarrow T + p \end{aligned} \qquad (2.10)$$

Fusion reactions among reaction products are not taken into account here. The net particle balance is therefore

$$4D \rightarrow {}^{3}He + T + n + p \qquad (2.11)$$

and the neutron yield amounts to 1/2 neutron per fusion reaction and 1/4 neutron per deuteron destroyed.

(b) SCAT-D fusion: (semicatalyzed deuterium fusion in which plasma operational conditions and ion density control are taken to be such that the bred tritium is burned at its rate of production and none of the ${}^{3}He$ is burned *in situ*):

$$\begin{aligned} D + D &\rightarrow T + p \\ D + T &\rightarrow {}^{4}He + n \\ D + D &\rightarrow {}^{3}He + n \\ \hline 5D &\rightarrow {}^{3}He + {}^{4}He + 2n + p \end{aligned} \qquad (2.12)$$

Here the neutron yield is 2/3 neutrons per fusion reaction and 2/5 neutrons per deuterium ion destroyed.

(c) CAT-D fusion: (catalyzed deuterium fusion in which conditions are assumed to be such that both the bred T and the bred ${}^{3}He$ are burned *in situ* at their rate of production):

$$\begin{aligned} D + D &\rightarrow T + p \\ D + T &\rightarrow {}^{4}He + n \\ D + D &\rightarrow {}^{3}He + n \\ D + {}^{3}He &\rightarrow {}^{4}He + p \\ \hline 6D &\rightarrow 2\,{}^{4}He + 2n + 2p \end{aligned} \qquad (2.13)$$

Here, the neutron yield is 1/2 neutron per fusion reaction and 1/3 neutron per deuterium ion destroyed.

TABLE 2.2
Tabulation of Various Fusion Modes and Associated Data

Fusion reactions (particle energies in parentheses, MeV)	Total energy released U_r (MeV)	Charged particle energy fraction	Neutron energy fraction
D–T burn			
$D + T \rightarrow n(14.1) + \alpha(3.5)$	17.6	0.20	0.80
D–D burn			
PURE-D			
$D + D \rightarrow T(1.0) + p(3.1)$	4.1		
$D + D \rightarrow {}^3He(0.8) + n(2.4)$	3.2		
$4D \rightarrow {}^3He + n + T + p$	7.3	0.67	0.33
SCAT-D			
$D + D \rightarrow T(1.0) + p(3.1)$	4.1		
$D + T \rightarrow n(14.1) + \alpha(3.5)$	17.6		
$D + D \rightarrow {}^3He(0.8) + n(2.4)$	3.2		
$5D \rightarrow {}^3He + 2n + \alpha + p$	24.9	0.34	0.66
CAT-D			
$D + D \rightarrow T(1.0) + p(3.1)$	4.1		
$D + T \rightarrow n(14.1) + \alpha(3.5)$	17.6		
$D + D \rightarrow {}^3He(0.8) + n(2.4)$	3.2		
$D + {}^3He \rightarrow \alpha(3.7) + p(14.6)$	18.3		
$6D \rightarrow 2n + 2\alpha + 2p$	43.2	0.62	0.38

Every D–D fusion mode is thus shown to produce neutrons, the difference being only in the extent of neutron production. We provide a detailed tabulation of these various fusion reactions and cycles, Table 2.2, listing the energies of the various reaction products as well as reaction and cycle yields for neutrons and fusion energy.

2.4. NEUTRON MULTIPLICATION

Neutron multiplication is defined as a neutron-induced process which results on average in more than one neutron emitted. One evident example is the fission reaction, Eq. (1.1). In addition, there are neutron-multiplying reactions in which the absorption of a neutron by a nucleus ${}^AZ$ results in x neutrons; the residual nucleus will generally be in an excited state and return to the ground state by gamma emission. We may characterize ie entire sequence as follows:

$$n + {}^{A}Z \rightarrow ({}^{A+1-x}Z)^{*} + xn$$
$$\searrow {}^{A+1-x}Z + \gamma \qquad (2.14)$$

Except for deuterium and beryllium, which have exceptionally low hreshold energies of ~3.3 and ~1.9 MeV, respectively, typical values for n, 2n) reactions are in the low MeV energy domain while the $(n, 3n)$ hreshold energies are somewhat higher.

.5. REACTION RATES

There are several interesting nucleonic features of a synergetic system equiring emphasis. One is the potentially larger range of neutron energies, the reason for which is that the source energies of neutrons appearing s a result of fusion or spallation reactions are generally well above the eak of a fission neutron spectrum (Figure 2.3). Another distinguishing eature is the important role of charged particle cross sections between arious light nuclei (nuclear fusion) and between light and heavy nuclei spallation).

In all instances microscopic cross sections are introduced to describe he probability for the occurrence of an r-type interaction between two articles. A geometric interpretation of the microscopic cross section, to e represented by σ_r, is that of an "effective size" of one reaction partner s seen by the other one; its dimensions are m^2 and the unit is the barn b), which is equal to 10^{-28} m^2.

As shown in Appendix B, the microscopic cross section σ_r must be nown for a calculation of reaction rates. The general presentation of the pace dependence of the rate at which an r-type reaction occurs between 4-type and B-type particles in a unit volume is

$$r_{r,AB}(\mathbf{x}) = \int_{\mathbf{v}_A}\int_{\mathbf{v}_B} v_{\text{rel}}\sigma_{r,AB}(v_{\text{rel}})\left[\frac{n_A(\mathbf{x})\,n_B(\mathbf{x})}{1+\delta_{AB}}\right] \times f_A(\mathbf{v}_A, \mathbf{x}) f_B(\mathbf{v}_B, \mathbf{x})\, d\mathbf{v}_B\, d\mathbf{v}_A \qquad (2.15)$$

where δ_{AB} is the Kronecker delta, the $\mathbf{v}_i$'s and n_i's refer to the particle velocity and the space-dependent particle number density of the two species A and B, respectively, and $f_i(\mathbf{v}_i, \mathbf{x})$ identifies the normalized velocity distribution function for $i = A, B$; v_{rel} is the relative speed

$$v_{\mathrm{rel}} = |\mathbf{v}_A - \mathbf{v}_B| \tag{2.16}$$

The total reaction rate is obtained from Eq. (2.15) by integration over the coordinate space $\mathbf{x}$:

$$R_{r,AB} = \int_V r_{r,AB}(\mathbf{x})\, d\mathbf{x} \tag{2.17}$$

Here the relevant domain of integration is the reaction volume V.

For the case that one particle, say B, is essentially at rest, it is common practice to express the reaction rate density as an integral with respect to the laboratory energy E_A of the moving particle A in the form

$$R_{r,AB} = \int_V n_B(\mathbf{x}) \int_{E_A} v(E_A)\sigma_{r,AB}(E_A) n_A(\mathbf{x}) f_A(E_A, \mathbf{x})\, dE_A\, d\mathbf{x} \tag{2.18}$$

Such an expression is of particular interest in a fission reactor with A representing a neutron and B a heavy, essentially stationary, nucleus.

This expression for the reaction rate can be further simplified by introducing the spectral number density of the moving particle in the coordinate-energy phase space by the definition

$$n(E_A, \mathbf{x}) = n_A(\mathbf{x}) f_A(E_A, \mathbf{x}) \tag{2.19}$$

Then, following the practice widely used in fission reactor physics, the space-dependent neutron flux is defined by

$$\phi(E_A, \mathbf{x}) = n(E_A, \mathbf{x}) v(E_A) \tag{2.20}$$

where $v(E_A) = \sqrt{E_A/2m_A}$ with m_A the particle mass and with the macroscopic cross section given by

$$\Sigma_{r,B}(E_A, \mathbf{x}) = n_B(\mathbf{x}) \sigma_{r,AB}(E_A) \tag{2.21}$$

then yields the familiar reaction rate expression

$$R_{AB} = \int_V \int_{E_A} \Sigma_{r,B}(E_A, \mathbf{x})\phi(E_A, \mathbf{x})\, dE_A\, d\mathbf{x} \tag{2.22}$$

The assumption that the B-type target nuclei are at rest in the laboratory system can be dropped without changing the form of the presentation as long as their velocity distribution is uncorrelated with that of the A-type nuclei. This yields Doppler broadened microscopic cross sections $\tilde{\sigma}$ with $\Sigma = n\tilde{\sigma}$.

For the case that both type of nuclei are in motion, as they are in thermonuclear fusion, the common procedure is to simplify Eq. (2.15) by defining a space-dependent reaction parameter

$$\langle \sigma v \rangle_{r,AB} = \int_{\mathbf{v}_A} \int_{\mathbf{v}_B} v_{\text{rel}} \sigma_{r,AB}(v_{\text{rel}}) f_A(\mathbf{v}_A, \mathbf{x}) f_B(\mathbf{v}_B, \mathbf{x})\, d\mathbf{v}_A\, d\mathbf{v}_B \tag{2.23}$$

for which the total reaction rate is then given in the form

$$R_{r,AB} = \int_V \frac{n_A(\mathbf{x}) n_B(\mathbf{x})}{1 + \delta_{AB}} \langle \sigma v \rangle_{r,AB}\, d\mathbf{x} \tag{2.24}$$

In a point-reactor model, the particle densities and the velocity distribution functions are taken to be space independent and the total fusion rate in the reaction volume V is then simply found to be

$$R_{r,AB} = \frac{n_A n_B}{1 + \delta_{AB}} \langle \sigma v \rangle_{r,AB} V \tag{2.25}$$

If the velocity distribution is taken to be Maxwellian and if one assumes $T_A = T_B = T$, then the fusion reaction parameter takes the form

$$\langle \sigma v \rangle = \left(\frac{\mu}{2\pi kT}\right)^{3/2} 4\pi \int_0^\infty v_{\text{rel}}^3 \sigma(v_{\text{rel}}) \exp\left(-\frac{\mu v_{\text{rel}}^2}{2kT}\right) dv_{\text{rel}} \tag{2.26}$$

where μ is the reduced mass.

A further discussion on this subject is found in Appendix B.

2.6. NUCLEAR CROSS SECTIONS

With the role of microscopic cross sections in the calculation of the reaction rates established, we consider those types of reactions which are of particular interest in a synergetic nuclear energy context. We concentrate first on the reactions introduced in the preceding section in which the target nucleus is essentially at rest. Examples are neutron-induced processes such as neutron absorption resulting in either radiative capture or fission for which

$$\sigma_a = \sigma_c + \sigma_{\text{fi}} \tag{2.27}$$

holds. The total microscopic cross section is then the sum of all partial cross sections, here

$$\sigma_t = \sigma_a + \sigma_s \tag{2.28}$$

with σ_s representing neutron scattering; these cross sections are dependent on the kinetic energy of the incident neutron in the laboratory system for a given medium temperature. All these cross sections are nucleus specific and can vary substantially with neutron energy.

The relative probability that a neutron will be captured, rather than causing a nucleus to fission, was introduced in Eqs. (2.2) and can now be expressed in terms of microscopic cross sections; we write for the parameter α of Eq. (2.3) specifically

$$\alpha = \sigma_c/\sigma_{\text{fi}} \tag{2.29}$$

The main features of the fission cross section for the important fissionable isotopes ^{233}U, ^{235}U, ^{238}U, ^{232}Th, and ^{239}Pu are displayed in Figure 2.7. The characteristics to note here are the increasing fission probability with decreasing neutron energy for the thermally fissile nuclei ^{233}U, ^{235}U, and ^{239}Pu and the threshold fission cross sections for ^{232}Th and ^{238}U. It is useful to note that these fission cross sections tend to become more alike in magnitude at high energies.

The breeding of thermally fissile nuclei is accomplished by neutron capture in ^{232}Th and ^{238}U, Eq. (1.2) and Eq. (1.3), while tritium fuel for the D–T fusion reaction is bred by neutron capture in lithium, Eq. (1.5) and Eq. (1.6). The neutron energy dependence of these breeding processes is displayed in Figure 2.8. The significant feature to note is the particularly effective tritium breeding by slow neutrons in ^{6}Li. The neutron-

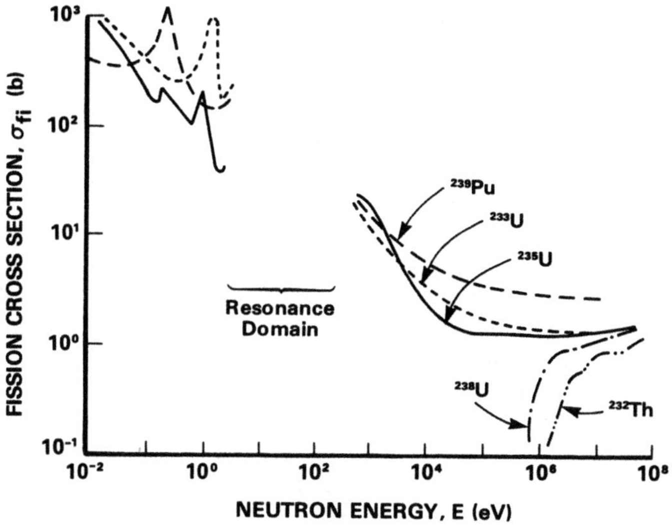

FIGURE 2.7. Fission cross section as a function of incident neutron energy.

conserving tritium breeding reaction in ^{7}Li has a threshold at about 2.8 MeV.

Another pertinent interaction example is the proton-induced spallation process. As suggested in Section 2.2, the accelerated protons need to possess energies well in the hundreds of MeV energy range in order to reduce the effect of Coulomb scattering interactions. The cross section

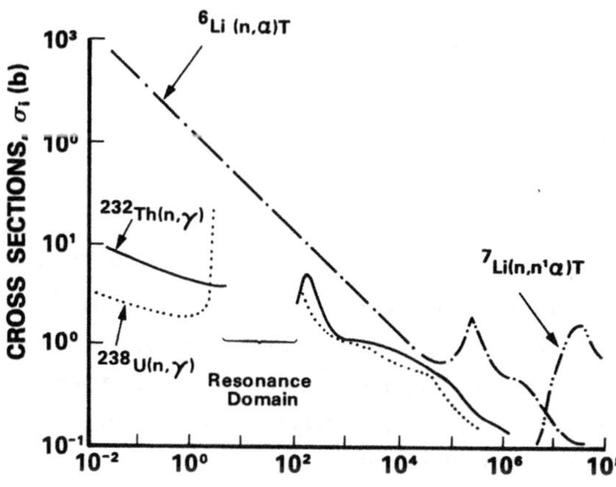

FIGURE 2.8. Capture cross sections for fissile and fusile fuel breeding.

for proton absorption depends on proton energy and may be approximately described by

$$\sigma_{sp} \simeq \begin{cases} 0, & E < a_8 B \\ \pi R_0^2 (1 - a_8 B/E), & a_8 B \leq E \end{cases} \tag{2.30}$$

and thus provides the conditions for a spallation reaction. Because of the electrostatic repulsion between the proton and the nucleus this reaction has an effective threshold at the proton energy $E = a_7 B$, where B is the height of the Coulomb barrier

$$B = \frac{Zq_e^2}{4\pi\varepsilon_0 R_0} \tag{2.31}$$

with the attenuation factor a_8 accounting for the quantum mechanical tunneling phenomenon; evidently

$$0 < a_8 < 1 \tag{2.32}$$

The symbol Z in Eq. (2.29) is the target proton number, q_e is the electron charge, and ε_0 is the permittivity of free space specified to give B in the same units as E. The quantity R_0 represents the empirical radius of the target nucleus, which is frequently taken to be

$$R_0 \simeq r_0 A^{1/3}, \qquad r_0 \simeq 1.25 \times 10^{-15}\,[\mathrm{m}] \tag{2.33}$$

except for very light nuclei.

When the nucleus spallates, it initially emits both protons and neutrons by direct cascade processes (Figure 2.4). These nucleons possess energies generally in the hundreds of MeV range. The emitted protons might induce further neutron-producing reactions or be slowed down to below the threshold energy $a_8 B$. The associated cascade neutrons, however, also possess a high probability of undergoing an inelastic reaction with the target nuclei. At these high energies, the pertinent absorption cross section is given approximately by

$$\sigma_a \simeq \pi(R_0 + \lambda)^2 \tag{2.34}$$

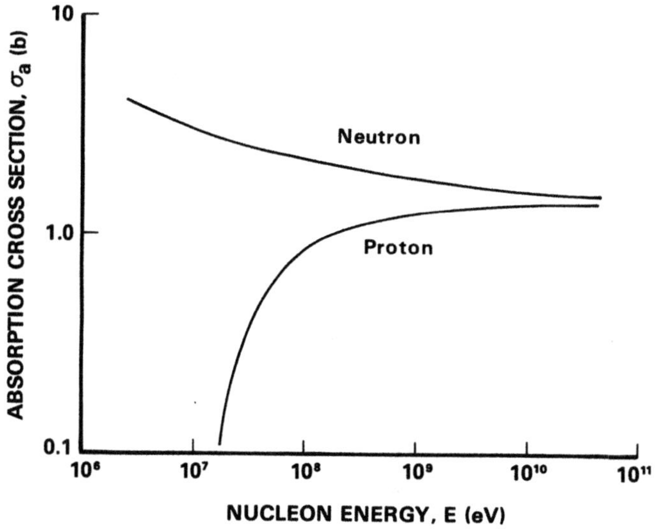

FIGURE 2.9. Depiction of high-energy neutron and proton absorption cross section as a function of nucleon energy.

where λ is the reduced de Broglie wavelength of the neutron,

$$\lambda \propto E^{-1/2} \tag{2.35}$$

with E as the neutron energy. Typical high-energy proton and high-energy neutron cross sections for medium-to-heavy nuclei are suggested in Figure 2.9.

Neutron multiplication cross sections are invariably threshold reactions requiring a minimum incident neutron energy generally in excess of 1 MeV. The notation for the neutron multiplication cross section is $\sigma_{n,xn}$, where $x = 2, 3, \ldots$; a set of such cross sections is schematically displayed in Figure 2.10. As a general characteristic, the (n, xn) cross sections first increase as E exceeds the threshold energy, reach a maximum, and then decrease as soon as E exceeds the threshold energy of the next-higher multiplication reaction $(n, (x + 1)n)$. Fission, of course, also represents a neutron multiplication process.

A common example of a reaction in which both reaction partners are in motion is thermonuclear fusion of light nuclei in a plasma, which, as discussed in the preceding section, may also be represented in terms of microscopic cross sections. Of interest here are those fusion processes which are most easily attained, implying that they can be realized with the lowest possible kinetic energies of the reacting particles. The three most easily achieved fusion reactions are D–T, D–D, and D–^{3}He. The fusion cross sections for these reactions are displayed in Figure 2.11.

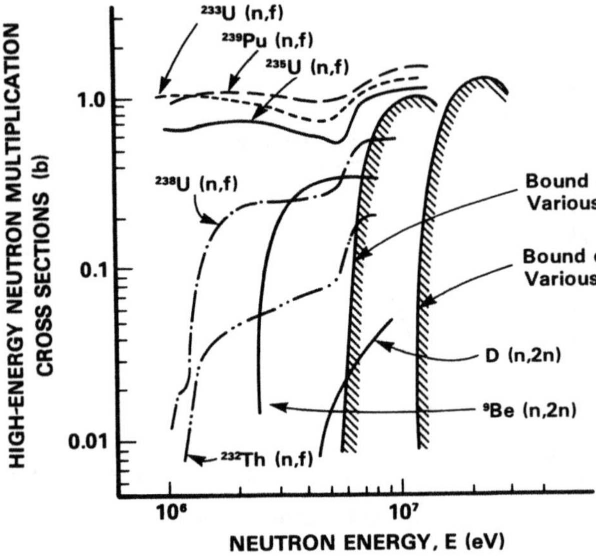

FIGURE 2.10. Neutron multiplication cross sections as a function of incident neutron energy.

Moreover, if we assume that the ions have an isotropic and space-independent velocity distribution which can be represented in terms of one parameter, we can plot the reaction parameter $\langle \sigma v \rangle$, Eq. (2.23), in terms of this one parameter. In the analysis of thermonuclear fusion, the velocity distribution of the ions is generally taken to be Maxwellian and

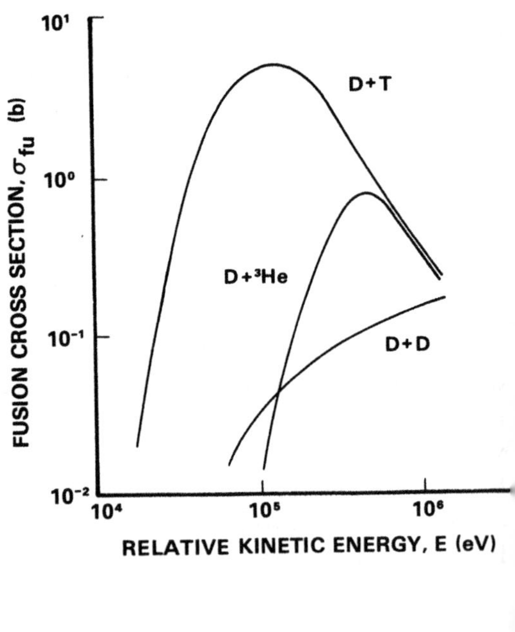

FIGURE 2.11. Fusion cross sections as a function of the relative energy of the fusion ions.

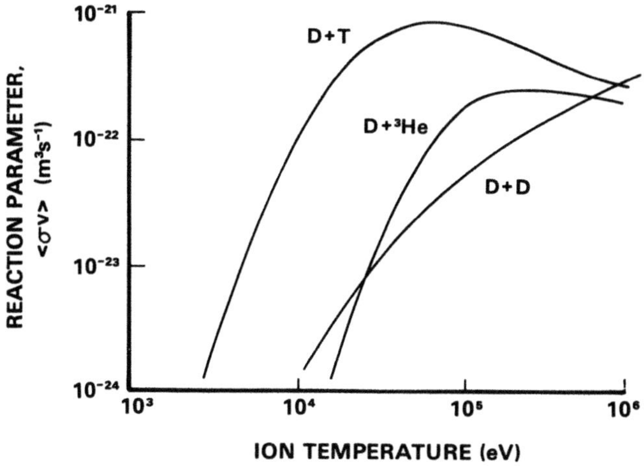

FIGURE 2.12. Fusion reaction parameter for a Maxwellian plasma.

is described in terms of the temperature of the ions. The reaction parameters $\langle \sigma v \rangle$, Eq. (2.26) for the above fusion reactions are shown in Figure 2.12 as a function in terms of the ion temperature kT, where k is the Boltzmann constant. For the case of a non-Maxwellian velocity distribution, the reaction parameters would have to be recalculated from Eq. (2.23) yielding reaction parameters different from those shown in Figure 2.12.

PROBLEMS

2.1. Determine the particle temperature for the mean-energy range of 1 eV to 1 MeV.

2.2. Examine the basis for the increasing similarity between neutron and proton absorption cross sections for energies in excess of 10^3 MeV.

2.3. Calculate for fission, fusion (D–T, PURE-D, SCAT-D, CAT-D), and spallation reactions the reaction energy associated with the production of one neutron, in units of MeV per neutron.

2.4. Formulate a general reaction balance analysis which leads to the estimation of threshold energies such as (n, xn) and ^{6}Li$(n, n'\alpha)$T.

2.5. Discuss the different role and the different importance of the energy distribution of the interacting particles (neutrons, protons, nuclei, ions) in fission and fusion, respectively. The discussion may be based on Eqs. (2.22) for fission and spallation, (2.24) for fusion, and on Figure 2.3.

2.6. Discuss the effect of neutron capture in fissile material (absorption without fission), Eq. (2.2), on the neutron and on the fissile fuel balance in a reactor. Specify for ^{233}U, ^{235}U, and ^{239}Pu by using a chart of isotopes which indicates the lifetimes and neutron absorption cross sections of the absorption products.

Chapter Three

REACTION AND REACTOR SYSTEMATICS

The introduction of alternative points of view or new concepts is invariably enhanced by the use of a notation which emphasizes the distinctiveness and unity of the subject. We introduce here a representation which is useful in describing general classes of nuclear processes as well as conventional and synergetic nuclear energy systems concepts. Our guide for the choice of notation is communicative clarity rather than completeness of physical detail.

3.1. REACTION NOTATION

The principal fission processes by thermal and fast neutrons involving thermally fissile nuclei of interest here can be written in the form

$$\left\{\begin{matrix} {}^{233}\mathrm{U} \\ {}^{235}\mathrm{U} \\ {}^{239}\mathrm{Pu} \\ {}^{241}\mathrm{Pu} \end{matrix}\right\} + n \rightarrow \nu_{\mathrm{Fi}} n + \sum_i \mathrm{Rp}_i \tag{3.1}$$

with Rp_i representing the associated fission reaction products. This equation may be written in two similar forms:

$$\mathrm{Fi}_t + n \rightarrow \nu_{\mathrm{Fi}} n + \sum \mathrm{Rp}_i \tag{3.2}$$

or, more compactly:

$$\mathrm{Fi}_t(n, \mathrm{Rp})\eta_{\mathrm{Fi}}n \tag{3.3}$$

Here Fi_t is any one of the thermally fissile nuclei ^{233}U, ^{235}U, ^{239}Pu, or ^{241}Pu and n is a neutron of arbitrary energy. Similarly, fast fission processes

$$\left\{\begin{matrix} ^{232}\mathrm{Th} \\ ^{238}\mathrm{U} \\ ^{240}\mathrm{Pu} \\ \vdots \end{matrix}\right\} + n_f \rightarrow \nu_{\mathrm{Fi}}n + \sum_i \mathrm{Rp}_i \tag{3.4}$$

can more generally be represented as

$$\mathrm{Fi}_f(n_f, \mathrm{Rp})\nu_{\mathrm{Fi}}n \tag{3.5}$$

where Fi_f denotes any one nucleus which will undergo fast fission induced by a fast neutron n_f. The general notation

$$\mathrm{Fi}(n, \mathrm{Rp})\nu_{\mathrm{Fi}}n \tag{3.6}$$

will therefore designate any of the above reactions with Fi being a fissionable, that is thermally or fast nucleus.

As discussed in Chapter 2, the reaction of relevance for nuclear fuel and neutron balance considerations is neutron absorption in fissionable nuclei and includes both fission and radiative capture reactions. Therefore we describe a fissile fuel destruction process by

$$\mathrm{Fi}(n, \mathrm{Rp})\eta_{\mathrm{Fi}}n \quad \text{or} \quad \mathrm{Fi}(n, \eta_{\mathrm{Fi}}n)\mathrm{Rp} \tag{3.7}$$

where Rp represents a heavy nucleus resulting from neutron capture or the two fission product nuclei for the case of neutron induced fission; here the net neutron yield is given by η_{Fi}. For completeness, then, the reaction

$$\mathrm{Fi}(n, \mathrm{Rp})\gamma \quad \text{or} \quad \mathrm{Fi}(n, \gamma)\mathrm{Rp} \tag{3.8}$$

describes the radiative neutron capture reaction where Rp represents the capture-product nucleus.

The above fission, absorption and capture reactions proceed in the reaction volume at the rate

$$R_{r,i} = \int_V \int_E \Sigma_{r,i}(E, \mathbf{x})\phi(E, \mathbf{x})\, dE\, d\mathbf{x} \tag{3.9}$$

where $\phi(E, \mathbf{x})$ is the space- and energy-dependent neutron flux and $\Sigma_{r,i}(E, \mathbf{x})$ is the macroscopic neutron cross section involving the i-type nuclei in a r-type reaction, Eqs. (2.20) and (2.21). Specifically, for fission and for radiative capture we write

$$\Sigma_{r,i}(E, \mathbf{x}) = n_i(\mathbf{x})\tilde{\sigma}_{r,i}(E), \qquad r = \text{fi}, c \tag{3.10}$$

and for absorption

$$\begin{aligned} \Sigma_{a,i}(E, \mathbf{x}) &= n_i(\mathbf{x})\tilde{\sigma}_{a,i}(E) \\ &= n_i(\mathbf{x})[\tilde{\sigma}_{c,i}(E) + \tilde{\sigma}_{\text{fi},i}(E)] \end{aligned} \tag{3.11}$$

Some of the relevant microscopic cross sections are displayed in Figures 2.7 and 2.8; $\tilde{\sigma}$ is discussed in Appendix B.

A general fusion reaction involving two fusile nuclei Fu and η_{Fu} fusion neutrons may be similarly written as

$$\text{Fu}(\text{Fu}, \text{Rp})\eta_{\text{Fu}}n \tag{3.12}$$

Examples for fusion reactions and fusion cycles are

$$\text{D}(\text{T}, {}^4\text{He})n \tag{3.13}$$

and

$$\text{D}(\text{D}, \text{Rp})\eta_{\text{Fu}}n \tag{3.14}$$

with η_{Fu} for the various deuterium fusion modes discussed in Section 2.3. For the above reactions and the assumption of a homogeneous plasma with a Maxwellian ion energy distribution we find, specifically for the D–T fusion reaction rate, the total fusion rate in the reaction volume V, as

$$R_{DT} = n_D n_T \langle \sigma v \rangle_{DT} V \tag{3.15}$$

and for D–D fusion, depending upon the fusion mode,

$$R_{Dj} = V \begin{cases} \frac{1}{2} n_D^2 \langle \sigma v \rangle_{DD}, & \text{PURE-D} \\ \frac{1}{2} n_D^2 \langle \sigma v \rangle_{DD} + n_D n_T \langle \sigma v \rangle_{DT}, & \text{SCAT-D} \\ \frac{1}{2} n_D^2 \langle \sigma v \rangle_{DD} + n_D n_T \langle \sigma v \rangle_{DT} + n_D n_{^3He} \langle \sigma v \rangle_{D^3He}, & \text{CAT-D} \end{cases} \tag{3.16}$$

In order to conform to our general space-independent or point-reactor model, the ion number densities n_i are assumed to be uniform throughout the reactor volume and $\langle \sigma v \rangle_{ij}$ are the reaction parameters as defined in Eq. (2.23) and displayed, for the special case of a Maxwellian ion distribution, in Figure 2.12. We reemphasize here our notation: n will always represent a neutron while n_i will always denote the density of the i-type ion.

The proton-induced spallation process can be represented in the form

$$p(Z, \mathrm{Rp}) \eta_{sp} n \tag{3.17}$$

where Z represents the high-Z target nucleus, Rp is the spallation product nucleus, and η_{sp} the net neutron yield per incident proton; this latter term differs from ν_{sp} of Eq. (1.16), which refers to neutron yield per spallation event. The rate at which such a spallation reaction proceeds in the entire target may be described by

$$\mathrm{R}_{sp} = \int_V \int_E \Sigma_{sp}(E, \mathbf{x}) \phi_p(E, \mathbf{x}) \, dE \, d\mathbf{x} \tag{3.18}$$

where E is the proton energy, $\phi_p(E, \mathbf{x})$ is the proton flux in the target, and $\Sigma_{sp}(E, \mathbf{x})$ the macroscopic spallation cross section

$$\Sigma_{sp}(E, \mathbf{x}) = \sum_i n_i(\mathbf{x}) \sigma_{sp,i}(E) \tag{3.19}$$

The breeding of thermally fissile and fusile fuel is represented as

$$\mathrm{Fb}(n, \beta\gamma)\mathrm{Fi} \tag{3.20}$$

and

$$\mathrm{Fb}(n, \mathrm{Rp})\mathrm{Fu} \tag{3.21}$$

where Fb is a fertile nucleus appropriate for breeding either fissile or fusile fuels, $\beta\gamma$ represents the beta and/or gamma radiations emitted as the transmutation proceeds following neutron capture, and Rp represents the reaction products which, in the case of T breeding from Li, are alpha particles and neutrons. The corresponding total reaction rate is

$$R_{c,\mathrm{Fb}} = \int_V \int_E \Sigma_{c,\mathrm{Fb}}(E, \mathbf{x})\phi(E, \mathbf{x})\, dE\, d\mathbf{x} \tag{3.22}$$

with $\Sigma_{c,\mathrm{Fb}}(E, \mathbf{x})$ as the macroscopic capture cross section for neutrons in the fertile material and $\phi(E, \mathbf{x})$ is the neutron flux.

3.2. REACTION LINKAGES

Reaction linkages can now be conveniently and compactly represented. For example, the fission chain may be written in general form simply as

$$\mathrm{Fi}(n, \mathrm{Rp})\eta_{\mathrm{Fi}}n \tag{3.23}$$

while the combined breeding and fission process is given by

$$\mathrm{Fb}(n, \beta\gamma)\mathrm{Fi}(n, \mathrm{Rp})\eta_{\mathrm{Fi}}n \tag{3.24}$$

This shorthand notation does not imply any particular rate at which these individual reactions occur.

The associated D–T fusion including tritium breeding is similarly written as

$$\mathrm{Fb}(n, \mathrm{Rp})\mathrm{Fu}(\mathrm{Fu}, \mathrm{Rp})\eta_{\mathrm{Fu}} n \tag{3.25}$$

Our use of the same symbol for a set of nuclei is not meant to overlook the important reaction and reactor physics characteristics of each of the individual isotopes. Thus, even though both ^{233}U and ^{239}Pu are represented by Fi, one could not simply exchange these two nuclides in a fission reactor without major reactor redesign. While a reactor needs to be designed and operated on a specific isotopic cycle, considerable generalization of the level of synergetic analysis is possible with the proposed notational set Fi, Fb, etc. We show in Appendix C how the concept of "equivalent fuel" may be incorporated in this generalized reactor physics analysis and will employ this approach in our subsequent study of specific symbiotic systems.

Furthermore, our emphasis on neutrons both as particles inducing reactions and as reaction products is not meant to imply that only neutrons are of interest. Take as an example the fusion chain reaction

$$\begin{aligned} p + {}^{6}\mathrm{Li} &\rightarrow {}^{3}\mathrm{He} + {}^{4}\mathrm{He} \\ {}^{6}\mathrm{Li} + {}^{3}\mathrm{He} &\rightarrow p + 2\,{}^{4}\mathrm{He} \end{aligned} \tag{3.26}$$

which can also be written as

$${}^{6}\mathrm{Li}(p, {}^{4}\mathrm{He}){}^{3}\mathrm{He}({}^{6}\mathrm{Li}, 2\,{}^{4}\mathrm{He})p \tag{3.27}$$

This reaction cycle has the same form as Eq. (3.25) with the proton p now being a chain carrier.

3.3. REACTOR REPRESENTATIONS

The notation introduced here can be conveniently extended to depict existing and projected synergetic nuclear power systems. For this purpose we introduce the following symbols:

$$[\;] \rightarrow \{\quad\}_X \begin{matrix} \rightarrow P_e \\ \rightarrow [\;] \end{matrix} \tag{3.28}$$

Here, the { } bracket is used to indicate the physical boundary of a nuclear power device or system as specified by X, and P_e is the nominal electric power output. The square bracket [ ] on the left side is used to indicate an external source of fuel—either fissile, fusile, or fertile—supplied to the nuclear device, and the same bracket on the right side refers to an eventual net output of fuel or other pertinent reaction products. The question of recovery of fissile fuels in the spent fuel rods will be commented upon subsequently. For the present we are concerned only with the net flow of power and of fuel.

As a first example illustrating the application of the above notation, we consider a fission burner reactor based on fuel highly enriched with ^{235}U such as is deployed in submarine and some research reactors. According to Eq. (3.28) this reactor can simply be described by the following:

$$[\mathrm{Fi}] \rightarrow \{\mathrm{Fi}(n, \mathrm{Rp})\eta_{\mathrm{Fi}}n\}_{\mathrm{BUR}} \rightarrow P_e \tag{3.29}$$

where the subscript BUR stands for burner reactor. The bracket [Fi] indicates that an external supply of fissile material is required. Some residual fissile conversion will invariably take place even in the case of burner reactors; however, because it occurs only to a relatively small extent we will not display it explicitly.

By an extension of the above, we obtain the graphical representation of the fission converter reactor, FCR, based on low ^{235}U enrichment:

$$\begin{matrix} [\mathrm{Fb}] \rightarrow \\ [\mathrm{Fi}] \rightarrow \end{matrix} \{\mathrm{Fb}(n, \beta\gamma)\mathrm{Fi}(n, \mathrm{Rp})\eta_{\mathrm{Fi}}n\}_{\mathrm{FCR}} \rightarrow P_e \tag{3.30}$$

The two important features to emphasize here are that both conversion and fission is taking place in the same core and that both fertile and fissile materials need to be supplied from external sources.

A fission breeder reactor, FBR, differs from Eq. (3.30) by the important change in the direction of net fissile fuel flow:

$$[\mathrm{Fb}] \rightarrow \{\mathrm{Fb}(n, \beta\gamma)\mathrm{Fi}(n, \mathrm{Rp})\eta_{\mathrm{Fi}}n\}_{\mathrm{FBR}} \begin{matrix} \rightarrow P_e \\ \rightarrow [\mathrm{Fi}] \end{matrix} \tag{3.31}$$

This reversal of the net flow of external fissile material means that it breeds sufficient fissile material to supply itself and provide for start-up or make-up fuel for other fission reactors.

Lastly, we define a fuel self-sufficient reactor SSR as a fission reactor which breeds exactly the same amount of fissile fuel it consumes:

$$[\mathrm{Fb}] \rightarrow \left\{\mathrm{Fb}(n, \beta\gamma)\mathrm{Fi}(n, \mathrm{Rp})\eta_{\mathrm{Fi}}n\right\}_{\mathrm{SSR}} \rightarrow P_e \tag{3.32}$$

The notation introduced here can be used to establish a convenient dimensionless characterization of the above four types of fission reactors. Recalling that the conversion ratio is defined as the average production rate of fissile nuclei divided by the average destruction rate of fissile nuclei, each of the above four reactor types is characterized by a conversion ratio given by

$$\frac{R_{c,\mathrm{Fb}}}{R_{a,\mathrm{Fi}}} = \begin{cases} = 0, & \text{fission burner reactor, BUR} \\ < 1, & \text{fission converter reactor, FCR} \\ = 1, & \text{fuel self-sufficient fission reactor, SSR} \\ > 1, & \text{fission breeder reactor, FBR} \end{cases} \tag{3.33}$$

Here, this ratio applies for equilibrium fuel cycle conditions.

The above fission reactor notation can be extended to fusion reactors. For the D–T fusion reactor, DTR, we use the following characterization

$$\begin{matrix}[\mathrm{D}] \rightarrow \\ [\mathrm{Li}] \rightarrow\end{matrix} \left\{\mathrm{Li}(n, \mathrm{Rp})\mathrm{T}(\mathrm{D}, {}^4\mathrm{He})n\right\}_{\mathrm{DTR}} \rightarrow P_e \tag{3.34}$$

where both Li and D need to be supplied from external sources.

The D–D fusion reactor, DDR, can be more simply represented by

$$[\mathrm{D}] \rightarrow \{\mathrm{D}(\mathrm{D}, \mathrm{Rp})\eta_{\mathrm{fu}}n\}_{\mathrm{DDR}} \rightarrow P_e \tag{3.35}$$

where, in this case of an isolated D–D fusion reactor, no use is made of the η_{Fu} fusion neutrons. Comparing this representation to the fission burner, Eq. (3.29), this kind of D–D fusion reactor may be classified as a "fusion burner." Aside from the role of fusion neutrons for fusion energy production, they are most valuable in a fission energy context. Hence, resource utilization provides the imperative for incorporating such a nuclear fusion reactor in a fission economy.

3.4. NUCLEAR SYMBIOSIS

The previously introduced reaction and reactor notation provide us with a convenient graphical tool to represent a variety of synergetic nuclear energy systems concepts. An important emphasis must be placed on the flow of isotopes and their transformation by nuclear-induced processes; beyond this, the focus is upon matching reaction products with reaction fuels involving various fusion, fission, and spallation processes. Though we include power flows in these graphical representations, detailed reaction and power analysis will be the subject of subsequent chapters.

As an initial example we take the spallation–fission system operating in a fuel self-sufficient mode. That is, we consider a spallation accelerator designed to breed fissile fuel in the blanket surrounding its target in order to provide the external fissile fuel required by one or several companion fission reactors. This concept represents the system suggested by Eq. (1.17) and Figure 1.2.

The spallation–fission symbiont incorporates the following basic system processes:

(a) proton-induced neutron production in the accelerator target, Eq. (3.17);
(b) neutron-induced fissile fuel breeding in the blanket surrounding the spallation target, Eq. (3.20);
(c) transfer of the bred fuel Fi to one or several associated fission converter reactors as represented by Eq. (3.29);
(d) fissile fuel consumption in the associated fission converter reactors, Eq. (2.2), Eq. (3.7).

Combining these components and using previously defined notation yields the following representation of the spallation–fission symbiont:

$$\begin{array}{lll} \begin{matrix}[Z] \to \\ [p] \to\end{matrix} & \left\{ p(Z, \mathrm{Rp})\,\eta_{\mathrm{sp}} n \right\}_{\mathrm{SpT}} & \rightleftarrows P_e \\ [\mathrm{Fb}] \to & \left\{ \mathrm{Fb}(n, \beta\gamma)\mathrm{Fi}(n, \mathrm{Rp})\,\eta_{\mathrm{Fi}} n \right\}_{\mathrm{SpB}} & \begin{matrix}\rightleftarrows P_e \\ \to [\mathrm{Fi}]\end{matrix} \\ \begin{matrix}[\mathrm{Fi}] \to \\ [\mathrm{Fb}] \to\end{matrix} & \left\{ \mathrm{Fb}(n, \beta\gamma)\mathrm{Fi}(n, \mathrm{Rp})\,\eta_{\mathrm{Fi}} n \right\}_{\mathrm{FCR}} & \begin{matrix}\to P_e \\ \to [\mathrm{Rp}]\end{matrix} \end{array} \quad (3.36)$$

Here the various rows refer to the system processes previously listed, SpT and SpB refer to the spallation target and the spallation blanket, respectively, and FCR identifies the companion fission converter reactor(s).

As indicated in Eq. (3.36) for the spallation–fission symbiont, only fertile materials, Fb, spallation target nuclei, Z, and the proton beam p, are externally supplied. The fission and activation products Rp generated as a by-product of neutron induced fission could possibly be used as target material and thus reduce the accumulation of radioactive nuclei.

The power produced in the target—made up of the deposited beam power and intrinsic nuclear power multiplication—may or may not be sufficient to compensate for the electric power required to accelerate the protons; here, we use the two-way arrow to indicate that the accelerator–target unit could be a net power consumer. In any case, the station as a whole must, of course, be a net producer of power.

With respect to the main source of power, the representation in Eq. (3.36) suggests that fission can occur in both the spallation blanket and the associated fission converter reactors. One may identify a variety of schemes, ranging between the two extreme cases of deliberately suppressing fission in the spallation blanket or of burning the bred fuel *in situ* in the spallation blanket. The first of these two limiting cases will be termed a spallation–fission symbiont which is, in particular, represented in Figure 1.2. The fission reactor is the dominant power producer while the spallation unit might produce only residual net power to the power grid or it may even be a net power consumer; its role in the symbiont context is mainly that of a fuel factory. Symbolically, we write:

$$\begin{array}{l}
\begin{array}{l}[Z]\rightarrow\\ [p]\rightarrow\end{array}\left\{p(Z,\mathrm{Rp})\eta_{\mathrm{sp}}n\right\}_{\mathrm{SpT}}\leftrightarrows P_e\\[2ex]
[\mathrm{Fb}]\rightarrow\left\{\mathrm{Fb}(n,\beta\gamma)\mathrm{Fi}\right\}_{\mathrm{SpB}}\rightleftarrows P_e \;\rightarrow[\mathrm{Fi}]\\[2ex]
\begin{array}{l}[\mathrm{Fi}]\rightarrow\\ [\mathrm{Fb}]\rightarrow\end{array}\left\{\mathrm{Fb}(n,\beta\gamma)\mathrm{Fi}(n,\mathrm{Rp})\eta_{\mathrm{Fi}}n\right\}_{\mathrm{FCR}}\rightarrow P_e
\end{array}\tag{3.37}$$

For the other extreme case all of the bred fuel is burned *in situ* making the spallation blanket the main power source:

$$\begin{array}{l}
\begin{array}{l}[Z]\rightarrow\\ [p]\rightarrow\end{array}\left\{p(Z,\mathrm{Rp})\eta_{\mathrm{sp}}n\right\}_{\mathrm{SpT}}\rightleftarrows P_e\\[2ex]
[\mathrm{Fb}]\rightarrow\left\{\mathrm{Fb}(n,\beta\gamma)\mathrm{Fi}(n,\mathrm{Rp})\eta_{\mathrm{Fi}}n\right\}_{\mathrm{SpB}}\rightarrow P_e
\end{array}\tag{3.38}$$

Such a system may be called a spallation–fission hybrid.

In the same vein, we may consider a D–T-fueled fusion reactor operating in symbiosis with a companion fission reactor or as a stand-alone hybrid. The distinguishing feature here is that the fusion-produced neutrons breed the required tritium as well as fissile fuel for a companion fission reactor. Using Eq. (3.12) specialized to D–T fusion, Eq. (3.13), and incorporating both tritium and fissile fuel breeding in the blanket, we represent the fusion–fission system operating in a fuel self-sufficient mode as follows:

$$
\begin{array}{ll}
[\mathrm{D}] \rightarrow \left\{ \mathrm{D}(\mathrm{T}, {}^4\mathrm{He})n \right\}_{\mathrm{DTC}} & \rightleftarrows P_e \\
[\mathrm{Li}] \rightarrow \left\{ \begin{array}{l} \mathrm{Li}(n, \mathrm{Rp})\mathrm{T} \\ \mathrm{Fb}(n, \beta\gamma)\mathrm{Fi}(n, \mathrm{Rp})\eta_{\mathrm{Fi}} n \end{array} \right\}_{\mathrm{DTB}} & \begin{array}{l} \rightleftarrows P_e \\ \leftrightarrow [\mathrm{Fi}] \end{array} \\
[\mathrm{Fb}] \rightarrow & \\
[\mathrm{Fi}] \rightarrow \left\{ \mathrm{Fb}(n, \beta\gamma)\mathrm{Fi}(n, \mathrm{Rp})\eta_{\mathrm{Fi}} n \right\}_{\mathrm{FCR}} & \rightarrow P_e \\
[\mathrm{Fb}] \rightarrow &
\end{array}
\tag{3.39}
$$

where DTC refers to the D–T reactor core and DTB refers to the D–T reactor blanket.

As suggested, the D–T fusion reactor itself could be a net power consumer. That is, the fusion reactor could very well operate below its energetic break-even condition while still allowing the overall system to be nonetheless energetically viable. This illustrates the feature that, whereas a fusion reactor operated as a stand-alone facility must be a net power producer, a fusion reactor combined with a fission reactor in a synergetically coupled context may be viable even as a net power consumer. The reason is that a fraction of the fission power is used to support the fusion neutron source which subsequently breeds fissile fuel much as fission power presently must indirectly supply power for uranium mining and the fissile fuel separation–enrichment plants.

One can now conceive of a number of novel and potentially effective variations of the above D–T-based fusion–fission concept. This can go even beyond the specialization of the general fusion–fission system, Eq. (3.39), to a fusion–fission symbiont or a fusion–fission hybrid by analogy to the spallation–fission system. For example, it may be judged worthwhile to design the companion fission reactor to breed tritium in its control rods and flux-flattening devices. In such a scheme, fission reactor control need not be affected but the neutrons used for such purposes now serve an additional function by becoming subsequently useful in the fusion–fission linkage. A vital characteristic of such concepts is that the breeding of tritium in a thermal fission reactor can evidently be accomplished very efficiently because of the $1/v$ cross section of ${}^6\mathrm{Li}$ (Figure 2.8), while concurrently permitting more of the 14-MeV fusion neutrons to be used for neutron multiplication (Figure 2.10). This would clearly aid neutron economy though it undoubtedly introduces design and operational complications.

The above fusion–fission systems are characterized by the need to supply lithium for tritium breeding. This feature of D–T fusion is eliminated by D–D fusion and, as shown in Table 2.2, also leads to neutron production to a varying extent. A D–D fusion reactor equipped with a fissile breeding blanket couples with a companion fission reactor in the following manner:

$$
\begin{array}{l}
[\mathrm{D}] \rightarrow \left\{ \mathrm{D(D, Rp)}\eta_{\mathrm{Fu}} n \right\}_{\mathrm{DDC}} \rightleftarrows P_e \\
[\mathrm{Fb}] \rightarrow \left\{ \mathrm{Fb}(n, \beta\gamma)\mathrm{Fi}(n, \mathrm{Rp})\eta_{\mathrm{Fi}} n \right\}_{\mathrm{DDB}} \begin{array}{l} \rightleftarrows P_e \\ \rightarrow [\mathrm{Fi}] \end{array} \\
\begin{array}{l} [\mathrm{Fi}] \rightarrow \\ [\mathrm{Fb}] \rightarrow \end{array} \left\{ \mathrm{Fb}(n, \beta\gamma)\mathrm{Fi}(n, \mathrm{Rp})\eta_{\mathrm{Fi}} n \right\}_{\mathrm{FCR}} \rightarrow P_e
\end{array}
\tag{3.40}
$$

The symbols DDC and DDB identify the D–D fusion reactor core and D–D fusion reactor blanket, respectively, with the D–D fusion mode dependent fusion neutron yield η_{Fu} used. This D–D-fueled system may be compared to its equivalent D–T system, Eq. (3.39).

Again a deuterium symbiont and hybrid mode of system operation can be envisaged. We represent the D–D fusion fission hybrid as

$$
\begin{array}{l}
[\mathrm{D}] \rightarrow \left\{ \mathrm{D(D, Rp)}\eta_{\mathrm{Fu}} n \right\}_{\mathrm{DDC}} \rightleftarrows P_e \\
[\mathrm{Fb}] \rightarrow \left\{ \mathrm{Fb}(n, \beta\gamma)\mathrm{Fi}(n, \mathrm{Rp})\eta_{\mathrm{Fi}} n \right\}_{\mathrm{DDB}} \rightarrow P_e
\end{array}
\tag{3.41}
$$

while the D–D fusion–fission symbiont follows by analogy.

As a last example of fusion–fission synergetic systems possibilities, we introduce the satellite reactor concept. By this we imply a stand-alone or coupled synergetic nuclear energy system supplying fuel for a number of smaller specialized electricity- or heat-producing reactors. As a case in point we describe here a satellite system based on D–^{3}He fusion.

As indicated in Chapter 2, one of the D–D fusion reaction channels leads to ^{3}He. Then, ^{3}He could fuse with D yielding charged particles as the only reaction products:

$$\mathrm{D} + {}^3\mathrm{He} \rightarrow {}^4\mathrm{He} + p \tag{3.42}$$

If concurrent D–D fusion reactions were minimized by appropriate reactor design and operation then this fusion process could well take place in separate decentralized specially designed small fusion reactors. These reactors could exploit the feature that only charged particles are emitted therefore making it possible to largely bypass the conventional thermodynamic cycle in favor of the more efficient direct conversion processes. Indeed, these reactors could potentially be radiologically relatively clean and therefore possess considerable siting advantages.

The supply of ^{3}He for the D–^{3}He satellites comes from the D–D fusion reactors based on the SCAT-D fusion, Eq. (2.12). This reactor could be operated in synergism with a companion fission reactor with the entire system represented as follows:

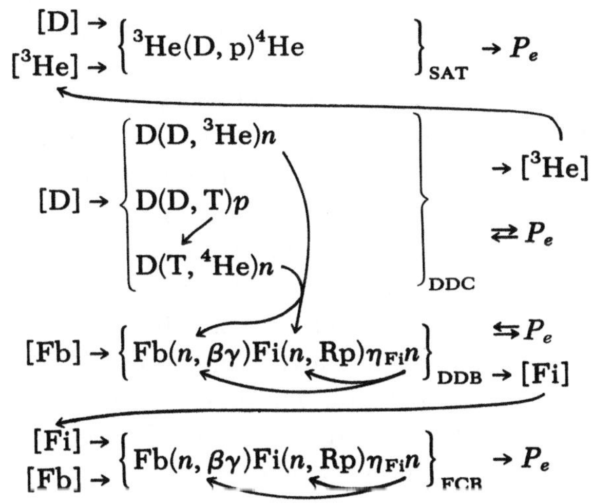

(3.43)

The first { } brackets represent the D–^{3}He satellite reactor, SAT, which produces power by direct conversion of the alpha and proton particles. The second { } bracket identifies the parent D–D fusion reactor core, DDC, which supplies ^{3}He to the satellite reactor and neutrons to the blanket, DDB, surrounding the fusion core. Finally, the fissile fuel bred in the blanket of the D–D reactor provides the companion fission reactor with all of its required fissile fuel.

As a variation on the above, the companion fission reactor could be replaced by a large number of small decentralized fission energy heat sources consisting of small self-regulating reactors. These small satellites could thus be regionally distributed to meet local demands for space heating. We address this concept in Chapter 9.

The above examples serve to indicate both the range and the kind of symbiotic nuclear energy systems which may be conceived. In each of these cases we have emphasized and incorporated the nuclear fuel supply aspect with the nuclear energy reaction processes; that is we have concentrated on the front-end aspects of the fuel cycle and nuclear energy processes. We now examine some aspects of the back-end of the fuel cycle which become apparent in a synergetic nuclear energy systems context.

3.5. BACK-END FUEL CYCLE

The problematic of the back end of the fuel cycle is largely associated with the radiological toxicity of the variety of radioactive isotopes associated with nuclear-energy-producing reactions. These radioisotopes can be grouped as follows:

(*a*) *Fission Products.* These radioactive nuclei have their origin in the initial product pairs of nuclear fission. The bimodal mass distribution is well known as are the half-lives and emitted radiations of the various isotopes and their daughter products. The radiological toxicity of spent fuel decreases approximately monotonically with time.

(*b*) *Actinides.* Actinides are isotopes formed as a result of successive neutron capture by heavy elements. They belong to the family of transuranic elements and are generally long-lived alpha-emitters.

(*c*) *Spallation Products.* Spallation products are the remains of proton-induced cascade and evaporation process. Their mass distribution is not well-known though theoretical considerations suggest a more uniform distribution than the fission product mass distribution.

(*d*) *Activation Products.* This group consists predominantly of those nuclei in structural, coolant, or other materials which have become radioactive as a result of reactions with neutrons.

While nuclear fuel can be produced from stable nuclei by nuclear transmutation, it is also in principle possible to render the radioactive nuclei of the back end of the fuel cycle stable, or of shorter half-life, by suitable nuclear transmutation processes. The conditions under which such incineration of radioactive wastes can occur is most important. The incorporation of fusion and spallation processes into a fission-based nuclear energy system provides some particularly interesting possibilities.

Since a fission reactor possesses a substantial inventory of unbound neutrons, it has long been thought that neutron capture should be an effective approach to transmuting fission products. For example, ^{137}Cs, which is one of the more plentiful fission products with a yield of about 0.06 per fission and a half-life of 30 years, could capture a neutron and therefore be transmuted as follows:

$$^{137}\mathrm{Cs} + n \xrightarrow{\gamma} {}^{138}\mathrm{Cs} \xrightarrow[32\ \mathrm{m}]{\bar{\beta}} {}^{138}\mathrm{Ba(stable)} \tag{3.44}$$

Evidently, this kind of incineration would have much appeal if it could be incorporated into a fission process according to

$$\begin{aligned} &\mathrm{Fi} + n \rightarrow \eta_{\mathrm{Fi}} n + \mathrm{Rp} \\ &\qquad n + \mathrm{Rp} + \cdots \rightarrow \mathrm{Rp_O} \end{aligned} \tag{3.45}$$

where Rp here is a radioactive fission or activation product and $\mathrm{Rp_O}$ is a stable nucleus obtained as a result of the neutron capture or, more generally, as the end product of a system of decay chain.

Coupling this concurrent incineration to fissile fuel breeding yields expressions of the form

$$\mathrm{Fb}(n, \beta\gamma)\mathrm{Fi}(n, \eta_{\mathrm{Fi}} n)\mathrm{Rp}(n, \ldots)\mathrm{Rp_O} \tag{3.46}$$

for a breeding–fission–incineration chain; here, as in Eq. (3.45), the dots represent the associated, if any, radiations or particles appropriate to the specific reaction. For D–T fusion taken as the neutron source, the incineration of Rp will appear as

$$\mathrm{Li}(n, n'\alpha)\mathrm{T}(\mathrm{D}, {}^4\mathrm{He})n(\mathrm{Rp}, \ldots)\mathrm{Rp_O} \tag{3.47}$$

where the fusion-neutron-induced reaction may, owing to the high-energy domain, include fast fission if Rp is a transuranic nucleus. The D–D incineration approach yields more simply

$$\mathrm{D}(\mathrm{D}, \mathrm{Rp})\eta_{\mathrm{Fu}} n(\mathrm{Rp}, \ldots)\mathrm{Rp_O} \tag{3.48}$$

A similar expression is provided by the neutron-rich spallation reaction

$$p + Z \rightarrow \eta_{sp} n + \sum_i \mathrm{Rp}_i$$
$$n + \mathrm{Rp} \rightarrow \mathrm{Rp}_0 \qquad (3.49)$$

or, in our shorthand notation

$$p(Z, \mathrm{Rp})\eta_{sp} n(\mathrm{Rp}, \ldots)\mathrm{Rp}_0 \qquad (3.50)$$

where Rp might be a radioactive species from any origin.

Several considerations enter into an assessment and implementation of neutron-based incineration, all of which invariably reduce to questions of neutron economy: the necessity for neutrons to close the fission chain translates to the need to breed fuel and to make up for the loss of neutrons by parasitic processes. The spectrum and magnitude of the neutron flux also enters as a parameter which can be illustrated by the following. The variation with time of the density of radioactive nuclei in a neutron field is given by the neutron capture rate and its natural decay rate

$$\frac{d}{dt} n_{\mathrm{Rp}}(t) = -\int_E \sigma_{c,\mathrm{Rp}}(E) n_{\mathrm{Rp}}(t)\phi(E)\, dE - \lambda_{\mathrm{Rp}} n_{\mathrm{Rp}}(t) \qquad (3.51)$$

where $n_{\mathrm{Rp}}(t)$ is the number density of the radioactive reaction product, $\sigma_{c,\mathrm{Rp}}(E)$ is the microscopic capture cross section, and λ_{Rp} is the radioactive decay constant of Rp. This latter parameter is related to its half-life $\tau_{1/2}$ by

$$\lambda_{\mathrm{Rp}} = \ln 2/\tau_{1/2} \qquad (3.52)$$

For the case of a suitably energy averaged expressions for the cross section and the neutron flux, Eq. (3.51) reduces to

$$\frac{d}{dt} n_{\mathrm{Rp}}(t) = -\sigma_{c,\mathrm{Rp}} n_{\mathrm{Rp}}(t)\phi - \lambda_{\mathrm{Rp}} n_{\mathrm{Rp}}(t) \qquad (3.53)$$

Evidently, a viable neutron-based incineration scheme should destroy the nuclear waste at a rate much faster than its natural radioactive decay rate. We require therefore

$$\sigma_{c,\mathrm{Rp}} n_{\mathrm{Rp}}(t)\phi \gg \lambda_{\mathrm{Rp}} n_{\mathrm{Rp}}(t) \tag{3.54}$$

More specifically, the neutron flux must satisfy

$$\phi \gg \frac{\lambda_{\mathrm{Rp}}}{\sigma_{c,\mathrm{Rp}}} \tag{3.55}$$

for the given isotope Rp. Substituting values for ^{137}Cs yields $\phi \gg 10^{19}\,(\mathrm{m}^{-2}\,\mathrm{s}^{-1})$, which is indeed a demanding requirement and generally not available.

In addition to particle balance considerations associated with nuclear incineration, it is necessary to also consider the overall energy balance. Clearly, less energy must be permitted to be consumed in the incineration of the radioisotope than was produced—on average—in the nuclear processes which yielded this nucleus.

Another feature to recognize is that neutron capture incineration is very wasteful of neutrons because of secondary neutron capture processes. This becomes clear by referring to Eq. (3.44) and noting that the radioactive intermediate product, ^{138}Cs, could capture a neutron before it beta decays to stable ^{138}Ba. Indeed, subsequent neutron captures expand the radioactive chains for ^{137}Cs as suggested in Figure 3.1.

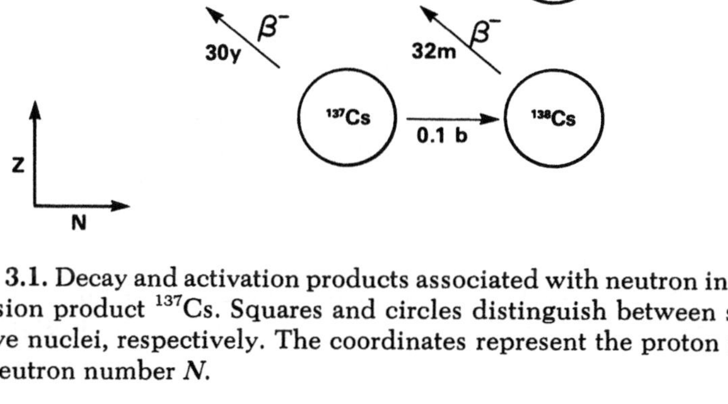

FIGURE 3.1. Decay and activation products associated with neutron incineration of the fission product ^{137}Cs. Squares and circles distinguish between stable and radioactive nuclei, respectively. The coordinates represent the proton number *Z* and the neutron number *N*.

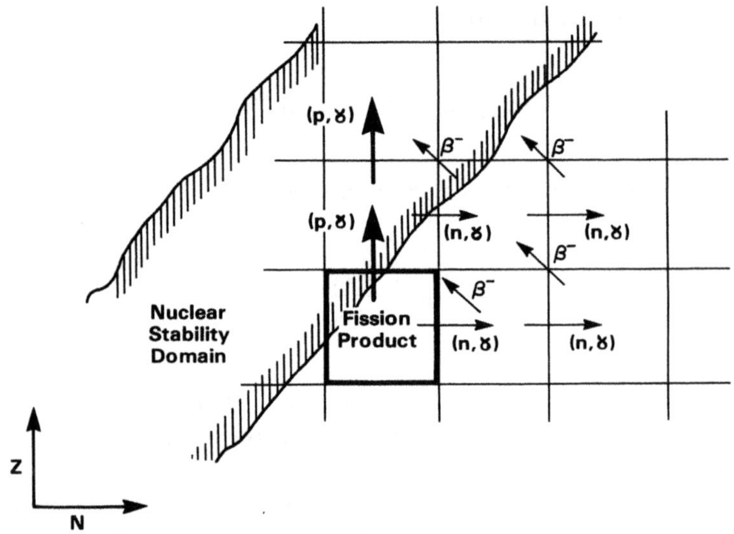

FIGURE 3.2. Schematic illustration showing the different reaction processes between neutron- and proton-induced transmutation of a fission product.

This general feature of successive neutron capture of fission products becomes evident from Figure 3.2. Fission products are normally neutron rich and thus tend towards the neutron surplus side of the nuclear stability domain. The likelihood of generating additional radioactive nuclei even farther from the line of stability is hence amplified for neutron-rich isotopes.

In the same figure, Figure 3.2, we also illustrate that proton capture transforms a radioactive fission product more directly into the stability domain. For ^{137}Cs, we have

$$^{137}\mathrm{Cs} + p \rightarrow {}^{138}\mathrm{Ba}\ (\text{stable}) \tag{3.56}$$

with no intermediate products. Unlike neutron capture, proton absorption requires that the proton possess sufficient energy to overcome Coulomb repulsion. Also the effect of Coulomb scattering and the pertinent energy loss of the proton may have a detrimental effect on the number of protons required per nucleus incinerated. Special proton accelerators could be considered provided that the secondary reactions do not significantly add to the toxicity burden.

Not all fission products can be transformed to a stable nucleus directly by proton capture. Taking ^{90}Sr, which is also a similarly high-yield fission product nuclei with a 29-year half-life, we obtain the following neutron and proton transmutation comparison:

$$^{90}\mathrm{Sr} + \begin{Bmatrix} n + {}^{91}\mathrm{Sr} \xrightarrow{\beta^-} \\ p \xrightarrow{\hspace{3em}} \end{Bmatrix} {}^{91}\mathrm{Y} \xrightarrow[59\ \mathrm{d}]{\beta^-} {}^{91}\mathrm{Zr}\ (\text{stable}) \tag{3.57}$$

Thus, in this case, proton transmutation bypasses one of the intermediate products.

3.6. FUEL REJUVENATION

As another back-end fuel cycle possibility we refer to what may be called refresh or rejuvenation of spent fuel via *in situ* reenrichment. We describe this in general by reference to a typical history of a fission reactor fuel pin for which the initial fissile enrichment is taken to be in the low enrichment range of 0.7%–3% leaving a 97% or more fertile inventory. As burn-up in the fuel pin—located in a fission reactor—takes place, the initial fissile inventory decreases and the bred fissile concentration produced by transmutation of the fertile nuclei builds up. Concurrently, fission products accumulate which, together with the net decrease of fissile nuclei, has the effect of reducing the reactivity contribution of the particular fuel pin. Eventually then, the fuel has to be withdrawn and replaced.

One may identify three types of options concerning the spent fuel pin. A direct option would involve its placement into permanent storage; this defines the so-called once-through fuel cycle. Another option would be to dissolve the fuel pin and extract the useful fissile material for incorporation into new fuel; this identifies the fissile fuel recycle option. The third may be termed *in situ* rejuvenation of the used fuel while it is retained in its cladding or, alternatively, reclad without prior dissolution of the fuel.

The fuel rejuvenation option makes use of the feature that an initially low-enrichment spent fuel, after removal from the reactor, still possesses a considerable amount of useful fertile material. If the entire fuel pin were now placed into an appropriate neutron field outside the reactor—such as a fusion or spallation blanket—and if the neutron spectrum at this location were carefully specified, then it is physically possible that sufficient fertile material may be converted into fissile fuel with a minimum of concurrent fissile nuclei destruction. The necessary condition for this *in situ* rejuvenation is that the ratio of neutron capture in the fertile nuclei to neutron absorption in fissile nuclei in the external nonfission neutron field, EXT, exceed by a significant margin the corresponding ratio in the fission converter reactor, FCR. Mathematically, this can be represented by

$$\left\{\frac{R_{c,\mathrm{Fb}}}{R_{a,\mathrm{Fi}}}\right\}_{\mathrm{EXT}} \gg \left\{\frac{R_{c,\mathrm{Fb}}}{R_{a,\mathrm{Fi}}}\right\}_{\mathrm{FCR}} \tag{3.58}$$

where $R_{c,\mathrm{Fb}}$ and $R_{a,\mathrm{Fi}}$ represent the neutron capture rates in the fertile nuclei and neutron absorption rates in the fissile nuclei, respectively, occurring in the fuel pin. The corresponding reaction rates are given by

$$R_{c,\mathrm{Fb}} = \sigma_{c,\mathrm{Fb}} n_{\mathrm{Fb}} \phi \tag{3.59}$$

and

$$R_{a,\mathrm{Fi}} = \sigma_{a,\mathrm{Fi}} n_{\mathrm{Fi}} \phi \tag{3.60}$$

where we have simplified the expressions in Eq. (3.9) to the case of a suitably energy averaged, homogeneous neutron field. The viability condition for rejuvenation, Eq. (3.58), is therefore met if the cross section ratios in the respective neutron environment fulfill the inequality

$$\left\{\frac{\sigma_{c,\mathrm{Fb}}}{\sigma_{a,\mathrm{Fi}}}\right\}_{\mathrm{EXT}} \gg \left\{\frac{\sigma_{c,\mathrm{Fb}}}{\sigma_{a,\mathrm{Fi}}}\right\}_{\mathrm{FCR}} \tag{3.61}$$

and if, in addition, the fertile-to-fissile fuel concentration is high enough:

$$\left\{\frac{\sigma_{c,\mathrm{Fb}}}{\sigma_{a,\mathrm{Fi}}} \cdot \frac{n_{\mathrm{Fb}}}{n_{\mathrm{Fi}}}\right\}_{\mathrm{EXT}} > 1 \tag{3.62}$$

We display in Figure 3.3 the capture-to-absorption cross-section ratios for the appropriate pairs of fertile–fissile nuclei. Recalling that the neutron energy in a thermal fission reactor peaks well below 1 eV, the above ratios are satisfied if the external neutron field possesses neutron spectra in the hundreds of keV domain and as long as the fissile enrichment of the fuel is low.

Calculational experimentation involving neutron fields in spallation accelerator and D–T fusion blankets confirm the conceptual validity of

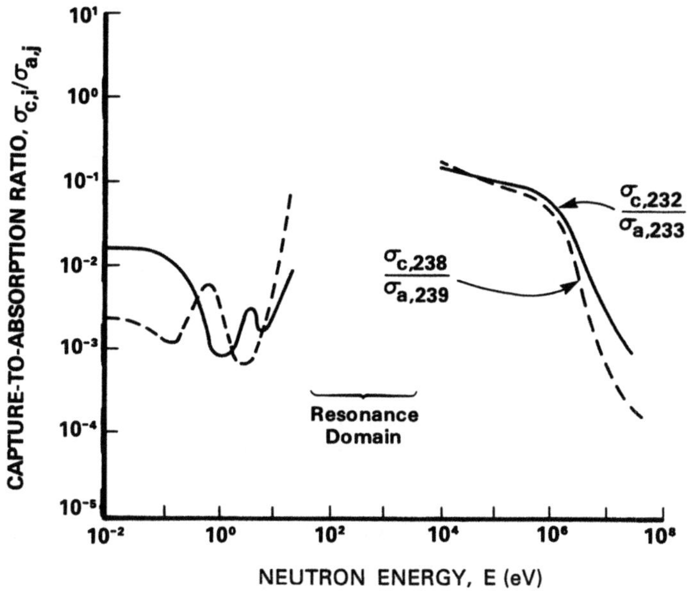

FIGURE 3.3. Capture-to-absorption cross-section ratios for pairs of fertile-to-fissile nuclei.

fuel rejuvenation, though it is anticipated that the fuel cladding may need to be replaced or redesigned in order to sustain the severe radiation environments. The number of such rejuvenation cycles would be limited by either radiation damage or *in situ* fertile fuel depletion.

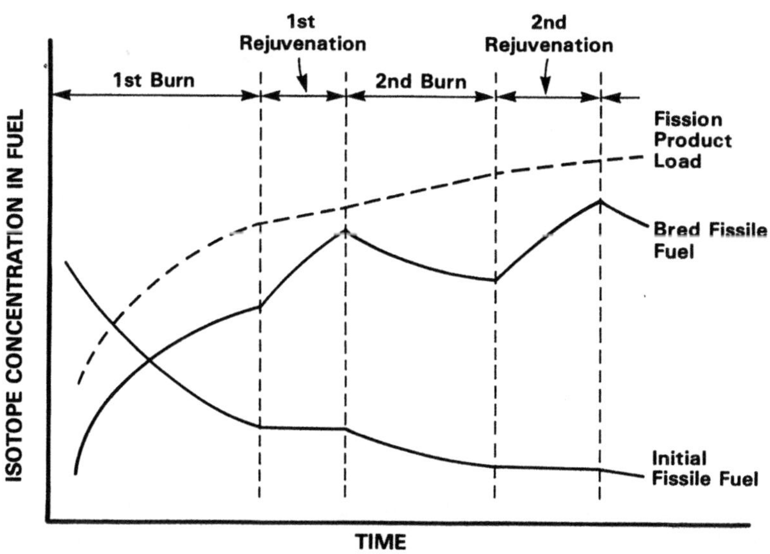

FIGURE 3.4. Schematic illustration of isotope concentration histories in the fuel as the fuel elements pass through a sequence of burn–rejuvenation cycles.

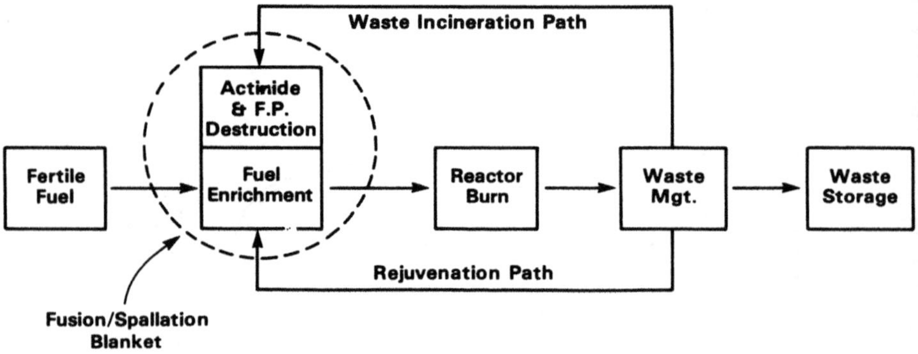

FIGURE 3.5. Block diagram representation of nuclear fuel cycle options which may be associated with a synergetic nuclear energy system.

On the basis of the above fuel rejuvenation considerations, typical isotopic concentrations histories in a fuel pin can be represented as suggested in Figure 3.4, where the initial fissile fuel and the bred fissile fuel are shown separately. Note that the initial fissile concentration decreases monotonically with time while that of the fission products increases; the bred fissile concentration increases during the rejuvenation periods in proportion to the neutron breeding reactions taking place in the external neutron fields. The concentration of the initial fissile fuel may change little during the rejuvenation period if little fission—and hence little power production—occurs by reason of the choice of neutron spectrum.

The above discussion on spent fuel rejuvenation and the preceding discussion of neutron- and proton-induced transmutation of radioactive waste nuclei suggests the potential for internal fuel renewal by suitable transfers of fuel between the various system elements. This is made feasible by the availability of nonfission neutron sources from fusion and spallation processes and the associated transmutation possibilities. We suggest in Figure 3.5 how such a fuel management and fuel replenishment might be represented for the case where only fertile material was externally supplied.

PROBLEMS

3.1. Nuclear synergism has the feature of integration in which the "output" from one part of the system provides the "input" to another. Identify examples of such synergism in ecology and biology.

3.2. Identify additional nuclear processes potentially useful in fission product transmutation.

.3. Identify one or several possibilities to improve the neutron balance of a fission reactor.

.4. Describe various processes which affect the length of successive burn–rejuvenation periods of a fuel element.

.5. Describe various processes which may compete, in fission nuclear waste transmutation, with the useful reactions. Also give an indication on how much energy may be invested in the destruction of a specific nucleus without jeopardizing the usefulness of fission as a power source.

.6. What are the technological, engineering, and operational characteristics which describe the relative merits of a hybrid and a symbiont, respectively?

Part II

SPALLATION–FISSION INTEGRATION

Chapter Four
THE SPALLATION–FISSION SYMBIONT

The production of neutrons by high-energy protons has been described in previous chapters. Our intent now is to establish an analytical framework for a synergetic nuclear energy system consisting of a spallation accelerator integrated with a companion fission reactor. In keeping with our chosen focus of analysis, we emphasize primarily the nuclear fuel and power balances.

4.1. SYSTEM DESCRIPTION

The concept of a spallation–fission symbiont has been introduced in Section 1.3. As suggested by Eq. (3.37) it consists of the following components and their specified functions:

(i) An accelerator produces high-energy protons which upon striking a target result in the release of neutrons available for fissile fuel breeding and other purposes; fission in the blanket surrounding the target is kept to a minimum.
(ii) A companion fission converter reactor receives the accelerator-bred fuel, which together with the fuel bred in the fission core, provides for its fissile fuel requirements.

Supplementary functions could involve the utilization of reactor-produced actinides as target material in the accelerator, fuel rejuvenation in the accelerator blanket, and destruction of selected fission products with spallation-produced nucleons. For reasons of separating functions, it is desirable that the blanket be free of the restrictive criticality constraints of a fission reactor core and to that end its effective reactivity state should be far below the critical level. Hence, in the combined spallation accelerator and fission converter reactor system, Eq. (3.36), there is reason to prefer that the spallation component be the main fuel producer while the fission reactor be the principal power producer.

In our initial analysis of the fuel balance, steady-state fuel self-sufficiency is imposed in the sense that only fertile fuel is externally supplied. For our analytical purposes we also impose the requirement that every fissile fuel nucleus appearing in the system will eventually be consumed in the system, implying therefore a perfectly closed and loss-free fuel cycle. These idealizations concerning the front end and the back end of the fuel cycle will be relaxed in subsequent sections.

There exists some latitude in the treatment of the power generated in the accelerator target–blanket assembly. If some of the bred fissile fuel were allowed to be retained in the blanket for an appreciable length of time then power production would occur and the extraction of this power for conversion and transmission to the power grid could be considered. Moreover, the beam power deposited in the target and further amplified by the nuclear processes therein could become a substantial contribution to the power balance. At the other extreme, it is conceivable that, for good reasons, the power produced in the target and blanket could be wasted.

A schematic representation of the main functions and associated processes of the spallation–fission symbiont is suggested in Figure 4.1.

4.2. NUCLEAR FUEL BALANCE

The fission reactor of this spallation–fission symbiont system destroys fissile fuel in order to produce nuclear power. In contrast, the principal function of the accelerator is to breed nuclear fuel. The reaction details suggested in Chapter 2 and the functional system and fuel flow representations given in Section 3.3 will now be used and expanded upon.

We consider first the fissile fuel balance in the fission converter reactor (FCR). Pertinent processes involving the fertile and fissile nuclei are represented by

$$\mathrm{Fb} + n \rightarrow \mathrm{Fi} + \cdots, \qquad R_{c,\mathrm{Fb}} \tag{4.1a}$$

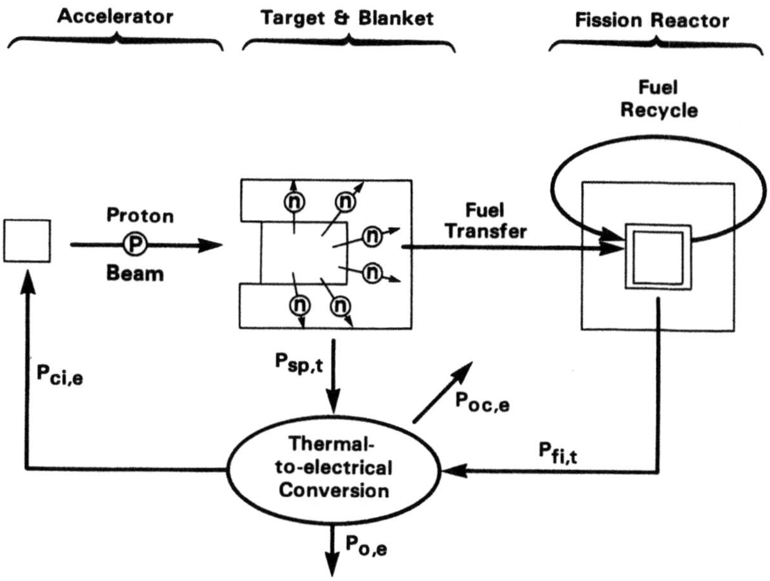

FIGURE 4.1. Depiction of fuel and power flow for the spallation–fission symbiont.

$$\mathrm{Fi} + n \rightarrow \eta_{\mathrm{Fi}} n + \sum_i \mathrm{Rp}_i, \qquad R_{a,\mathrm{Fi}} \tag{4.1b}$$

Recall that Fb and Fi designate a fertile and a fissile nucleus, respectively, and Rp represents a reaction product resulting from either a radiative capture or a fission process. The rate expression associated with the production and consumption of fissile fuel in the fission converter reactor at any time of reactor operation, is therefore described by

$$\left\{\frac{d}{dt} N_{\mathrm{Fi}}\right\}_{\mathrm{FCR}} = \{R_{c,\mathrm{Fb}} - R_{a,\mathrm{Fi}}\}_{\mathrm{FCR}} \tag{4.2}$$

where N_{Fi} is the fissile inventory in the fission conversion reactor, that is, the total number of fissile nuclei in the reaction volume at an arbitrary time. Secondary fissile production–destruction rates, such as production by (n, xn) reactions processes as well as by photofission, are not explicitly listed because their contribution is relatively small. Nevertheless, they could be incorporated either as specific terms or as fractional contributions to one of the two terms of Eq. (4.2).

In order to relate the two reaction rate expressions in Eq. (4.2) to the main absorption reaction rate, we introduce the space and energy averaged conversion ratio defined as

$$C_{\mathrm{Fi}} = \frac{R_{c,\mathrm{Fb}}}{R_{a,\mathrm{Fi}}}$$

$$= \frac{\int_V \int_E \Sigma_{c,\mathrm{Fb}}(E, \mathbf{x})\Phi(E, \mathbf{x})\, dE\, d\mathbf{x}}{\int_V \int_E [\Sigma_{\mathrm{fi,Fi}}(E, \mathbf{x}) + \Sigma_{c,\mathrm{Fi}}(E, \mathbf{x})]\Phi(E, \mathbf{x})\, dE\, d\mathbf{x}} \tag{4.3}$$

which is consistent with the earlier breeding capacity classification of fission reactors, Eq. (3.33). For converter fission reactors, C_{Fi} is in the range

$$0 < C_{\mathrm{Fi}} < 1 \tag{4.4}$$

In general, the various reaction rates and, consequently, the conversion ratio, are slow functions of time but for our purposes will be taken as time independent. That is, all quantities are taken as appropriately time averaged. In a later section this constraint will be relaxed for those parameters which have a pronounced time dependence.

Using the conversion ratio C_{Fi} to eliminate $R_{c,\mathrm{Fb}}$ in Eq. (4.2) gives

$$\left\{\frac{d}{dt} N_{\mathrm{Fi}}\right\}_{\mathrm{FCR}} = \{(C_{\mathrm{Fi}} - 1)R_{a,\mathrm{Fi}}\}_{\mathrm{FCR}} \tag{4.5}$$

Clearly, in a convertor reactor the factor $(C_{\mathrm{Fi}} - 1)$ is negative, implying a decrease in fissile core inventory with time.

The determination of the fissile breeding rate expression for the accelerator target–blanket assembly requires some additional considerations. Denoting the proton current as I_p amperes and the proton charge as q_p in units of Coulombs gives I_p/q_p beam protons per second entering the target.

We had previously indicated, Eq. (2.5) and Figure 2.5, that on average, each incident proton will yield a total of η_{sp} neutrons. A fraction of these are taken to be captured in fertile nuclei eventually yielding μ_{sp} fissile nuclei per neutron. Hence, the fissile fuel production rate in the spallation blanket, SpB, is given by

$$\left\{\frac{d}{dt} N_{\mathrm{Fi}}\right\}_{\mathrm{SpB}} = \{\mu_{\mathrm{sp}}\eta_{\mathrm{sp}}I_p/q_p\}_{\mathrm{SpB}}$$

$$= \{C_p I_p/q_p\}_{\mathrm{SpB}} \tag{4.6}$$

where

$$C_p = \mu_{sp}\eta_{sp} \tag{4.7}$$

is the number of fissile fuel nuclei bred per incident beam proton.

The various appropriate accelerator and target–blanket properties are represented in this expression: the beam current is given by I_p, proton energy and the target size and compositions determine η_{sp}, and target–blanket composition and geometry is incorporated in μ_{sp}. A more detailed analysis would require several additional phenomena to be included. For example, some of the fissile nuclei bred in the spallation blanket may undergo fission before they are removed from the blanket and the fuel breeding chains may not go to completion by reason of neutron capture in the intermediate isotopes. Hence, effective efficiency factors could be included in Eq. (4.6) as multipliers. Rather than including such additional factors we include these effects in the definition of C_p making it, appropriately, the net number of fissile nuclei supplied by the spallation blanket to the associated fission reactor per incident proton.

Eqs. (4.5) and (4.6) constitute the complete descriptions—for our model assumptions—of the fissile fuel production/consumption processes for the spallation symbiont. The net fissile fuel balance at any time is therefore given by

$$\left\{\frac{d}{dt}N_{\text{Fi}}\right\}_o = \left\{\frac{d}{dt}N_{\text{Fi}}\right\}_{\text{FCR}} + \left\{\frac{d}{dt}N_{\text{Fi}}\right\}_{\text{SpB}}$$

$$= \{(C_{\text{Fi}} - 1)R_{a,\text{Fi}}\}_{\text{FCR}} + \{C_p I_p/q_p\}_{\text{SpB}} \tag{4.8}$$

Here and in the following, the subscript o refers to an overall net parameter. If $\{dN_{\text{fi}}/dt\}_o$ is positive then fissile fuel is accumulating and available for other fission reactors; if it is negative then fissile fuel from other sources needs to be supplied.

4.3. FUEL SELF-SUFFICIENCY

Systems parameters such as fission reactor power, proton beam current, and breeding rates in the blanket under steady-state operational conditions are evidently important in the initial assessment of a facility. In order to assess the relative role of some of these systems parameters we introduce an important design–operational criterion: on average, at any time after start-up, the spallation–fission symbiont is to be self-sufficient with respect to its fissile fuel requirements. This implies that, once the initial fissile inventory to the companion fission reactor has been supplied, there exists no need to provide further fissile fuel from external sources and, additionally, no excess fissile fuel will be produced. As previously emphasized, all bred fissile nuclei are assumed to eventually fission so that no special analytical fuel cycle management schemes need here be introduced. In terms of Eq. (4.8) then, the overall fuel self-sufficiency requirement—at an arbitrary instant in time—is defined by

$$\left\{\frac{d}{dt} N_{\mathrm{Fi}}\right\}_o = 0 \tag{4.9}$$

By Eq. (4.8) fission reactor parameters C_{Fi} and $R_{a,\mathrm{Fi}}$ and spallation accelerator parameters C_p, I_p are then related by

$$\{(1 - C_{\mathrm{Fi}})R_{a,\mathrm{Fi}}\}_{\mathrm{FCR}} = \{C_p I_p / q_p\}_{\mathrm{SpB}} \tag{4.10}$$

The parameters for a spallation–fission symbiont under the imposed fuel self-sufficiency conditions are thus specified. This coupling condition assures the fuel self-sufficient mode of operation and can be interpreted as the condition to be imposed on the spallation component in order to have it support one or several fission converter reactors, or, alternatively, for a given spallation component, the number of fission reactors which can be supported as characterized by $R_{a,\mathrm{Fi}}$ and the associated fission power.

However, Eq. (4.10) is not the only pertinent constraint condition. The total station electrical output will invariably be another design criterion of major importance. We therefore consider the station power balance and from this then obtain a relationship which will specifically involve Eq. (4.10).

4.4. POWER BALANCE

As depicted by Figure 4.1, the overall net electrical power supplied to the grid, $P_{o,e}$, is given in terms of the following power components:

$$P_{o,e} = \eta_{\mathrm{fi},g} P_{\mathrm{fi},t} + \eta_{\mathrm{sp},g} P_{\mathrm{sp},t} - P_{\mathrm{ci},e} - P_{oc,e} \tag{4.11}$$

where the first two terms represent the various gross electric power production contributions and the last two terms represent internally consumed power components. Specifically, the power terms $P_{\mathrm{fi},t}$ and $P_{\mathrm{sp},t}$ represent the thermal power generated in the fission converter reactor and in the spallation accelerator target and blanket, respectively; the factors $\eta_{\mathrm{fi},g}$ and $\eta_{\mathrm{sp},g}$ are the respective thermal-to-electrical conversion efficiencies relating the thermal station output to the gross electric station output for each component.

The electric power $P_{oc,e}$ required to operate and control the entire system may be taken to be a fraction of the thermal power contributions

$$P_{oc,e} = \varepsilon_{\mathrm{fi}} P_{\mathrm{fi},t} + \varepsilon_{\mathrm{sp}} P_{\mathrm{sp},t} \tag{4.12}$$

and may include energy in the form of electricity or steam. Introducing Eq. (4.12) into Eq. (4.11) yields

$$P_{o,e} = (\eta_{\mathrm{fi},g} - \varepsilon_{\mathrm{fi}}) P_{\mathrm{fi},t} + (\eta_{\mathrm{sp},g} - \varepsilon_{\mathrm{sp}}) P_{\mathrm{sp},t} - P_{ci,e} \tag{4.13}$$

This suggests the introduction of net thermal-to-electric conversion efficiencies η_{fi} and η_{sp} which implicitly incorporate operation and control requirements for the entire system consisting of the fission reactor, the spallation target, and the blanket facility. The station power output may be more compactly written as

$$P_{o,e} = \eta_{\mathrm{fi}} P_{\mathrm{fi},t} + \eta_{\mathrm{sp}} P_{\mathrm{sp},t} - P_{ci,e} \tag{4.14}$$

The thermal fission reactor power can be represented in terms of the fissile fuel consumption rate $R_{a,\mathrm{Fi}}$, Eq. (3.9)

$$P_{\mathrm{fi},t} = U_{\mathrm{Fi}} R_{a,\mathrm{Fi}} \tag{4.15}$$

where U_{Fi} is the total recoverable energy released per fissile nucleus consumed. Actually, Eq. (4.15) may be taken as the defining equation for U_{Fi} which includes the net contributions of the various reactions taking place in the fission reactor. Hence, U_{Fi} can be approximately related to the fission reaction energy $U_{r,\mathrm{Fi}}$ introduced in Section 1.1 according to

$$U_{\mathrm{Fi}} R_{a,\mathrm{Fi}} \simeq U_{r,\mathrm{Fi}} R_{\mathrm{fi},\mathrm{Fi}} + \sum_{i \neq \mathrm{Fi}} \sum_{j} U_{r_j,i} R_{j,i} \tag{4.16}$$

Here the first term on the right side represents the contribution from fissioning of the fissile fuel nuclei while the second term includes processes involving other nuclei. Since the reaction energy $U_{r,\mathrm{Fi}}$ is approximately the same for all heavy nuclei (Table 2.1), we find

$$U_{\mathrm{Fi}} \simeq \frac{1+f}{1+\alpha} U_{r,\mathrm{Fi}} \tag{4.17}$$

where $(1+f)$ is the factor by which fast fission of nonfissile nuclei increases the energy yield per fissile nucleus consumed, and $1/(1+\alpha)$ is the factor by which radioactive capture in fissile fuel decreases this energy contribution. For the various types of fission converter reactors, these two factors combine to give very approximately unity, so that

$$U_{\mathrm{Fi}} \simeq U_{r,\mathrm{Fi}} \simeq 190 \text{ MeV} \tag{4.18}$$

With $R_{a,\mathrm{Fi}}$ in units of inverse seconds and U_{Fi} in MeV, the power $P_{\mathrm{fi},t}$ in Eq. (4.15) possesses units of MeV/s, which can be converted to any other convenient unit of power.

For the spallation target–blanket power we write similarly

$$P_{\mathrm{sp},t} = U_{\mathrm{sp}} I_p / q_p \tag{4.19}$$

where I_p/q_p is the number of beam protons entering the target per second and U_{sp} is the average energy released per incident proton. However, unlike the fission energy release, U_{sp} depends strongly on the proton energy, target composition, and target geometry as alluded to in Figure 2.6. Blanket effects will be considered in Section 4.6.

The term $P_{\mathrm{ci},e}$ in Eq. (4.11) refers to the internally circulating electrical power required to produce the proton beam. Using η_p as the efficiency of converting electrical power into proton beam power, P_p, we obtain for the circulating electric power component the expression

$$P_{\mathrm{ci},e} = P_p / \eta_p \tag{4.20}$$

The electrical power delivered to the grid may therefore be represented in terms of the more fundamental reactor and accelerator parameters by substitution of Eqs. (4.15), (4.19), and (4.20) into Eq. (4.14) to yield

$$P_{o,e} = \eta_{\mathrm{fi}} U_{\mathrm{Fi}} R_{a,\mathrm{Fi}} + \eta_{\mathrm{sp}} U_{\mathrm{sp}} I_p / q_p - P_p / \eta_p \tag{4.21}$$

where we have assumed that the waste heat deposited in the accelerator, $(1-\eta_p)P_{\mathrm{ci},e}$, is not recovered.

A more useful expression is one which shows specifically the two dominant power constituents: thermal fission power of the fission reactor and beam power of the accelerator. This prompts us to relate the beam power to the beam current. With I_p/q_p protons entering the target per second and each proton possessing an energy E_p, the proton beam power P_p is

$$\begin{aligned} P_p &= E_p I_p/q_p \\ &= \frac{P_{\mathrm{sp},t}}{U_{\mathrm{sp}}/E_p} \end{aligned} \tag{4.22}$$

where use of Eq. (4.19) has been made. It is evident that the ratio U_{sp}/E_p appearing in Eq. (4.22) represents the intrinsic energy gain of the target configuration, suggesting therefore that we define the spallation target energy gain parameter, G_{sp} of Figure 2.6, by

$$G_{\mathrm{sp}} = U_{\mathrm{sp}}/E_p \tag{4.23}$$

Inserting Eqs. (4.22) and (4.23) into Eq. (4.21) gives the desired expression for the net electric power output of the spallation–fission symbiont in terms of the net electric power of the fission reactor and of the accelerator beam power:

$$\begin{aligned} P_{o,e} &= \eta_{\mathrm{fi}} U_{\mathrm{Fi}} R_{a,\mathrm{Fi}} + [\eta_{\mathrm{Sp}} G_{\mathrm{sp}} - 1/\eta_p] P_p \\ &= \eta_{\mathrm{fi},o} P_{\mathrm{fi},t} + \eta_{\mathrm{sp},o} P_{\mathrm{sp},t} \end{aligned} \tag{4.24}$$

Here we have introduced the overall thermal-to-electrical conversion efficiencies for the fission component, $\eta_{\mathrm{fi},o}$, and the spallation component, $\eta_{\mathrm{sp},o}$. These overall efficiencies of the components are apparently functions of the net conversion efficiencies and the fractional power recycling requirement

$$\begin{aligned} \eta_{\mathrm{fi},o} &= \eta_{\mathrm{fi}} \\ &= \eta_{\mathrm{fi},g} - \varepsilon_{\mathrm{fi}} \end{aligned} \tag{4.25}$$

and

$$\eta_{sp,o} = \eta_{sp} - \frac{1}{G_{sp}\eta_p}$$

$$= \eta_{sp,g} - \varepsilon_{sp} - \frac{1}{G_{sp}\eta_p} \tag{4.26}$$

One may expect that generally $\eta_{fi,g} \gg \varepsilon_{fi}$ so that $\eta_{fi,o}$ is positive. If $\eta_{sp,o}$ is negative then the spallation component is a burden on the power balance. This would of course be acceptable if the power deficiency were not excessive with respect to the power output of the associated fission reactor since the accelerator blanket provides fuel and thus replaces conventional power-consuming fissile fuel mining and processing operations. For $\eta_{sp,o}$ negative, a suitable requirement is that the net energy consumption of the spallation component be small compared to the energy output of the fission component

$$|\eta_{sp,o}| \ll \eta_{fi}\left(\frac{P_{fi,t}}{P_p G_{sp}}\right) \tag{4.27}$$

in order to tend towards enhanced energy viability of the spallation–fission symbiont. If $\eta_{sp,o} = 0$, then the accelerator component may be called self-powered and produces power just to meet its own requirements.

With the general objective of optimizing power conversion, Eq. (4.24) provides two parametric relationships useful to guide system design:

$$\text{Max: } \{\eta_{fi,g} - \varepsilon_{fi}\} \tag{4.28}$$

and

$$\text{Max: } \left\{\eta_{sp,g} - \varepsilon_{sp} - \frac{1}{\eta_p G_{sp}}\right\} \tag{4.29}$$

These would clearly be major design objectives for a spallation–fission system.

We now consider the total station electrical power output in terms of the companion fission reactor thermal power for the special case of fuel self-sufficiency. Rearranging Eq. (4.10) we obtain the relationship

$$\frac{I_p}{q_p} = \left(\frac{1 - C_{\mathrm{Fi}}}{C_p}\right) R_{a,\mathrm{Fi}} \tag{4.30}$$

Introducing the fission reactor power, Eq. (4.15), as well as the beam power, Eq. (4.22), yields

$$\left(\frac{1 - C_{\mathrm{Fi}}}{U_{\mathrm{Fi}}}\right) P_{\mathrm{fi},t} = \left(\frac{C_p}{E_p}\right) P_p \tag{4.31}$$

which can be used to define the fuel self-sufficiency fission-to-beam power ratio in terms of appropriate physical and design parameters

$$\frac{P_{\mathrm{fi},t}}{P_p} = \left(\frac{U_{\mathrm{Fi}}}{E_p}\right)\left(\frac{C_p}{1 - C_{\mathrm{Fi}}}\right) \tag{4.32}$$

Substitution of Eq. (4.32) in Eq. (4.24) yields an expression for the net electric power output of this fuel self-sufficient spallation–fission system in terms of the thermal fission reactor power:

$$P_{o,o} = \left[\eta_{\mathrm{fi},o} + \eta_{\mathrm{sp},o}\left(\frac{U_{\mathrm{sp}}}{U_{\mathrm{Fi}}}\right)\left(\frac{1 - C_{\mathrm{Fi}}}{C_p}\right)\right] P_{\mathrm{fi},t} \tag{4.33}$$

This is a useful and compact expression for the electric power delivered to the grid by a spallation–fission symbiont operating in a fuel self-sufficient mode.

Note further that this symbiont concept has meaning only for $C_{\mathrm{Fi}} < 1$ since otherwise there will be no need to supply fuel to the companion fission reactor. This condition is clearly evident in Eq. (4.32) implying that $P_p \to 0$ as $C_{\mathrm{Fi}} \to 1$ as would be expected. For conversion ratios in the neighbourhood of unity, which characterize high-converter and near-breeder reactors, the fissile fuel requirement of the fission component would indeed be very small; in such a case, the initial fission reactor inventories—not included in our definition of fuel self-sufficiency—would have a major impact on the overall fissile fuel requirements.

4.6. FISSION AND BEAM POWER

A numerical assessment of the fission power to beam power ratio imposed on a spallation–fission symbiont by the fuel self-sufficiency mode of operation requires specification of various constants or their ratios. We take Eq. (4.32) as the expression of interest and assume the fission reactor to conform to the present generation of reactors represented by $C_{\mathrm{Fi}} \sim 0.5$ and we use $U_{\mathrm{fi}} \simeq 190$ MeV. For the companion spallation component, the parameters cannot be taken from actual devices and have to be extrapolated from related experiments, calculations, and expectations for further developments. For the beam particle energy we take $E_p \simeq 1000$ MeV and suppose that the target is of medium size and consists of fissionable material; from Figure 2.5 we then find $\eta_{\mathrm{sp}} \simeq 53$. Assuming arbitrarily that each of the spallation neutrons yields 0.95 externally available fissile nuclei $\mu_{\mathrm{sp}} = 0.95$, gives $C_p \simeq 50$. The fission-to-beam power ratio, Eq. (4.32) is then

$$P_{\mathrm{fi},t}/P_p \simeq 19 \tag{4.34}$$

Therefore, a 3000-MW$_t$ companion fission reactor could be associated with a spallation accelerator whose proton beam power is $P_p \simeq 160$ MW. Further, if this beam power is obtained from electrical power with a conversion efficiency of $\eta_p \sim 0.5$ then some 320 MW$_e$ needs to be supplied. The power balance thus becomes very important since a 3000-MW$_t$ fission reactor delivers some 1000 MW$_e$. Note that the ratio of thermal fission component power to thermal spallation component power, $P_{\mathrm{fi},t}/P_{\mathrm{sp},t}$, can be directly calculated from Eq. (4.34) if the spallation energy gain $G_{\mathrm{sp}} = U_{\mathrm{sp}}/E_p$ is known. By extension, Figure 2.6 suggests that this ratio is in the interval $0.5 \lessapprox G_{\mathrm{sp}} \lessapprox 8$ and clearly is a function of the design of the target.

Additional insight can be obtained from a parametric assessment in which the conversion ratio C_{Fi} of the fission reactor component and the breeding capacity of the spallation component, as given by the net number of fissile nuclei supplied per incident proton $C_p = \mu_{\mathrm{sp}}\eta_{\mathrm{sp}}$, are taken as variables.

The accelerator current for the above beam power requirement can be readily calculated from Eq. (4.22):

$$I_p = (q_p/E_p)P_p \tag{4.35}$$

If the beam power is taken in units of MW, the proton energy is in units of MeV, and the proton charge is expressed in Coulombs, we obtain the convenient numerical relation

$$I_p[\mathrm{A}] = \frac{P_p[\mathrm{MW}]}{E_p[\mathrm{MeV}]} \tag{4.36}$$

where the unit of each symbol is indicated. Together with Eq. (4.30) this yields

$$I_p[\mathrm{A}] = \left(\frac{P_{\mathrm{fi,t}}[\mathrm{MW}]}{U_{\mathrm{Fi}}[\mathrm{MeV}]}\right)\left(\frac{1-C_{\mathrm{Fi}}}{C_p}\right) \tag{4.37}$$

For a system consisting of a current 3000-MW_t fission reactor and a spallation accelerator with a 1000-MeV proton beam operating in a fuel self-sufficient mode we find

$$I_p \simeq 0.16\,[\mathrm{A}] = 160\,[\mathrm{mA}] \tag{4.38}$$

for the average proton current.

Reducing the adverse effect of the potentially high accelerator power requirements can be accomplished by two different approaches. One would be to design the spallation blanket as an efficient energy multiplier and fully use the resultant energy in the power balance. This may have its limitations because in so doing one has to increase the fission rate in the blanket, which would tend to transform it into a subcritical fission reactor driven by externally produced spallation neutrons. This may be incorporated in our previous formulation by defining a spallation blanket power multiplication factor M_{SpB} which incorporates the additional power released by fission reactions induced in the blanket. In this case Eq. (4.19) would become

$$P_{\mathrm{sp},t} = M_{\mathrm{SpB}}U_{\mathrm{sp}}I_p/q_p \tag{4.39}$$

Another approach is to choose a neutronically more efficient companion fission reactor. For example, if in the above example the reactor had a conversion ratio of 0.9 rather than 0.5 with all other parameters remaining unchanged, the proton beam power would be reduced to 30 MW_e.

These considerations suggest a useful truism of nuclear energy synergism: it is most important to carefully choose the partners for an effective symbiotic relationship.

4.7. ENRICHED FUEL SUPPLY

Throughout the preceding analysis of the spallation symbiont we have assumed that, except for the initial start-up fissile inventory in the fission reactor, all externally supplied fuels consisted only of fertile atoms. The fission fuel requirement of the companion fission reactor was taken to be met by the instantaneous breeding capacity of the accelerator blanket and the recycle provisions associated with the fission reactor conversion process. No consideration was given to the isotopic concentration of fissile nuclei in the fuel, that is to the enrichment of the fuel, which is invariably one of the design parameters of a fission reactor.

Our intent now is to include fissile enrichment into our fuel flow considerations. To do so we consider the spallation blanket—or an arbitrary part of it—as an isotopic control volume and undertake a fissile and fertile isotope accounting. Furthermore we intend to consider the effect of rejuvenating spent fuel, or of supplying low-enrichment fuel to the blanket surrounding the accelerator target where it is enriched to the level required by the fission reactor; the low-enriched fuel supply could be natural or depleted uranium, with the latter being supplied from diffusion plant stockpiles.

The enrichment of a fuel is defined by

$$\varepsilon = \frac{N_{\mathrm{Fi}}}{N_{\mathrm{Fi}} + N_{\mathrm{Fb}}} \tag{4.40}$$

where N_{Fi} and N_{Fb} are the number of fissile and fertile atoms, respectively in a given fuel volume. In particular we designate by ε_R the enrichment required by the fission converter reactor and by ε_X the enrichment of an external fuel feed stock from which fertile ($\varepsilon_X = 0$) or marginally enriched fuel ($\varepsilon_X < \varepsilon_R$) is supplied to the spallation blanket.

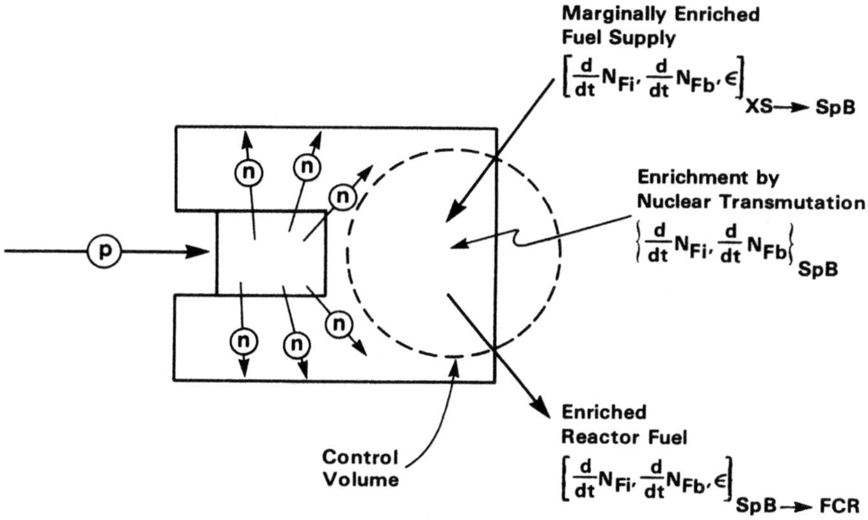

FIGURE 4.2. Schematic of fuel flow in the spallation blanket.

We will find it useful to extend the nuclear fuel balance considerations of Section 4.2 and consider the isotope flow schematic suggested by Figure 4.2. Note in particular the characterization of the fuel supplied to the accelerator blanket, the implied enrichment process in the control volume, and the subsequent outflow destined for the companion fission reactor.

We again employ the steady-state operational mode previously defined and impose therefore the continuity requirement on the specified fuel mass containing both fissile and fertile nuclei:

$$\left[\frac{d}{dt}N_{Fi}\right]_{XS \rightarrow SpB} + \left\{\frac{d}{dt}N_{Fi}\right\}_{SpB} - \left[\frac{d}{dt}N_{Fi}\right]_{SpB \rightarrow FCR} = 0 \quad (4.41)$$

and

$$\left[\frac{d}{dt}N_{Fb}\right]_{XS \rightarrow SpB} + \left\{\frac{d}{dt}N_{Fb}\right\}_{SpB} - \left[\frac{d}{dt}N_{Fb}\right]_{SpB \rightarrow FCR} = 0 \quad (4.42)$$

In both these equations the first term represents the supply from an external stock, XS, while the second term stands for the net fissile fuel production or fertile fuel destruction rate, respectively, in the spallation blanket (SpB); the third term is the net amount of the fissile and fertile fuel in the reactor-grade fuel transferred to the fission converter reactor (FCR). Evidently, the first term in Eq. (4.41) and Eq. (4.42) represents the extension of our previous analysis, Eq. (4.1). Note also that the symbolic notation of Eq. (4.41) and Eq. (4.42) represents the reactor representation introduced in Section 3.3. To distinguish between transfer of fuel to and from, and production/destruction of fuel in a system component, we use [ ] brackets for the former and { } brackets for the latter.

Accounting for neutron capture and fission of both the fissile and fertile nuclei in the spallation blanket yields the following rate equations in terms of the previously defined rate expressions:

$$\left[\frac{d}{dt}N_{\mathrm{Fi}}\right]_{\mathrm{XS}\to\mathrm{SpB}} + \{R_{c,\mathrm{Fb}} - R_{c,\mathrm{Fi}} - R_{\mathrm{fi,Fi}}\}_{\mathrm{SpB}} - \left[\frac{d}{dt}N_{\mathrm{Fi}}\right]_{\mathrm{SpB}\to\mathrm{FCR}} = 0 \quad \textbf{(4.43)}$$

and

$$\left[\frac{d}{dt}N_{\mathrm{Fb}}\right]_{\mathrm{XS}\to\mathrm{SpB}} + \{-R_{\mathrm{fi,Fb}} - R_{c,\mathrm{Fb}}\}_{\mathrm{SpB}} - \left[\frac{d}{dt}N_{\mathrm{Fb}}\right]_{\mathrm{SpB}\to\mathrm{FCR}} = 0 \quad \textbf{(4.44)}$$

These two equations include reaction processes not enumerated in the previous more simplified analysis, Eq. (4.2). Any comparison with the preceding analysis will therefore require imposing the same reaction conditions.

We will use Eqs. (4.43) and (4.44) as expressions for which the reaction rates are variables and require therefore elimination of the differential terms. For this purpose we introduce the enrichments ε_R and ε_X for the fuel mass moving through the system, which will permit elimination of two of the four time derivatives. We rewrite Eq. (4.40) in the form

$$N_{\mathrm{Fb}} = \left(\frac{1-\varepsilon}{\varepsilon}\right) N_{\mathrm{Fi}} \quad \textbf{(4.45)}$$

and apply this relation to the total fuel mass consisting of both the marginally enriched fuel supplied to the blanket, ε_X, and the reactor-grade enriched fuel destined for the fission reactor, ε_R. This allows us to eliminate N_{Fb} in Eq. (4.44) and yields the following set of two equations in two unknown fissile fuel time derivatives:

$$\left[\frac{d}{dt}N_{\mathrm{Fi}}\right]_{\mathrm{XS}\to\mathrm{SpB}} - \left[\frac{d}{dt}N_{\mathrm{Fi}}\right]_{\mathrm{SpB}\to\mathrm{FCR}} = \{R_{a,\mathrm{Fi}} - R_{c,\mathrm{Fb}}\}_{\mathrm{SpB}} \tag{4.46}$$

and

$$\frac{1-\varepsilon_X}{\varepsilon_X}\left[\frac{d}{dt}N_{\mathrm{Fi}}\right]_{\mathrm{XS}\to\mathrm{SpB}} - \frac{1-\varepsilon_R}{\varepsilon_R}\left[\frac{d}{dt}N_{\mathrm{Fi}}\right]_{\mathrm{SpB}\to\mathrm{FCR}} = \{R_{a,\mathrm{Fb}}\}_{\mathrm{SpB}} \tag{4.47}$$

The supply rate of fissile materials from external stockpiles, $[dN_{\mathrm{Fi}}/dt]_{\mathrm{XS}\to\mathrm{SpB}}$, can now be conveniently eliminated to yield one equation connecting the enriched fissile fuel supply rate, $[dN_{\mathrm{Fi}}/dt]_{\mathrm{SpB}\to\mathrm{FCR}}$, with the several essential parameters of the spallation–fission system:

$$\left[\frac{d}{dt}N_{\mathrm{Fi}}\right]_{\mathrm{SpB}\to\mathrm{FCR}}\left(\frac{1-\varepsilon_X}{\varepsilon_X} - \frac{1-\varepsilon_R}{\varepsilon_R}\right) = \left\{R_{a,\mathrm{Fb}} + \frac{1-\varepsilon_X}{\varepsilon_X}(R_{c,\mathrm{Fb}} - R_{a,\mathrm{Fi}})\right\}_{\mathrm{SpB}} \tag{4.48}$$

To use this and other expressions, we return to our previous discussion leading to Eq. (4.9) in which the net fissile fuel balance equation for the steady-state fuel self-sufficiency mode of operation was considered. For the present case we write

$$\left\{\frac{d}{dt}N_{\mathrm{Fi}}\right\}_o = \left\{\frac{d}{dt}N_{\mathrm{Fi}}\right\}_{\mathrm{FCR}} + \left[\frac{d}{dt}N_{\mathrm{Fi}}\right]_{\mathrm{SpB}\to\mathrm{FCR}} = 0 \tag{4.49}$$

The expression for $\{dN_{\mathrm{Fi}}/dt\}_{\mathrm{FCR}}$, Eq. (4.5), used previously may again be employed. In contrast, the term $[dN_{\mathrm{Fi}}/dt]_{\mathrm{SpB}\to\mathrm{FCR}}$ in this equation—which serves the same function as $\{dN_{\mathrm{Fi}}/dt\}_{\mathrm{SpB}}$ in the analysis of Section 4.2, Eq. (4.8)—is now given for the enriched case by Eq. (4.48). Substituting for Eq. (4.5) and Eq. (4.48) in Eq. (4.49) therefore provides conditions on the spallation symbiont to ensure that it operates at steady-state with the arbitrary enrichments of ε_X for the input fuel and ε_R for the output fuel:

$$\{(1-C_{\mathrm{Fi}})R_{a,\mathrm{Fi}}\}_{\mathrm{FCR}} = \frac{\varepsilon_R}{\varepsilon_R-\varepsilon_X}\left\{R_{c,\mathrm{Fb}}\left[1 + \varepsilon_X\frac{R_{\mathrm{fi,Fb}}}{R_{c,\mathrm{Fb}}} - (1-\varepsilon_X)\frac{R_{a,\mathrm{Fi}}}{R_{c,\mathrm{Fb}}}\right]\right\}_{\mathrm{SpB}} \tag{4.50}$$

According to our initial analysis which led to Eq. (4.7), we may use

$$\{R_{c,\mathrm{Fb}}\}_{\mathrm{SpB}} = C_p I_p / q_p \tag{4.51}$$

and rewrite the right-hand side of the self-sufficiency condition for a spallation–fission symbiont, Eq. (4.51), in the form

$$\{(1 - C_{\mathrm{Fi}})R_{a,\mathrm{Fi}}\}_{\mathrm{FCR}} = \frac{\varepsilon_R}{\varepsilon_R - \varepsilon_X}\left\{C_p \frac{I_p}{q_p}\left[1 + \varepsilon_X \frac{R_{\mathrm{fi,Fb}}}{R_{c,\mathrm{Fb}}} - (1 - \varepsilon_X)\frac{R_{a,\mathrm{Fi}}}{R_{c,\mathrm{Fb}}}\right]\right\}_{\mathrm{SpB}} \tag{4.52}$$

It is useful to compare this equation with Eq. (4.10), particularly in the limit $\varepsilon_X \to 0$.

4.8. EXTENDED FUEL FLOW

For the specified assumption of a loss-free, completely closed fuel cycle in which each fissile nucleus is eventually consumed, Eq. (4.52) represents a comprehensive formulation of the instantaneous parameter requirements for a spallation–fission symbiont operating in steady state with an external fuel supply characterized by the supply enrichment ε_X. Evidently, Eq. (4.52) represents a rigid coupling of the fission reactor parameters on the left-hand side of the equation with the spallation/accelerator parameters on the right-hand side. When compared to the results previously obtained, Eq. (4.10), two additional aspects appear. One is related to the external supply of preenriched fuel, $\varepsilon_X > 0$, which modifies the spallation breeding requirement essentially by the factor $\varepsilon_R/(\varepsilon_R - \varepsilon_X)$ while the other is related to proper fertile and fissile fission accounting.

In this context, the term $R_{\mathrm{fi,Fb}}/R_{c,\mathrm{Fb}}$ is formally defined by

$$\left\{\frac{R_{\mathrm{fi,Fb}}}{R_{c,\mathrm{Fb}}}\right\}_{\mathrm{SpB}} = \frac{\int_{\mathrm{SpB}} n_{\mathrm{Fb}}(\mathbf{x}) \int_E \sigma_{\mathrm{fi,Fb}}(E)\phi(E,\mathbf{x})\, dE\, d\mathbf{x}}{\int_{\mathrm{SpB}} n_{\mathrm{Fb}}(\mathbf{x}) \int_E \sigma_{c,\mathrm{Fb}}(E)\phi(E,\mathbf{x})\, dE\, d\mathbf{x}} \tag{4.53}$$

where n_{Fb} is the fertile atom number density. It is therefore largely determined by the inverse capture-to-fission ratio of the fertile nuclei, Eq. (2.29).

The latter term on the right-hand side of Eq. (4.52) describes the difference between the amount of fissile fuel bred in the spallation blanket and the amount which eventually becomes available for the companion fission reactor. This term

$$\left\{\frac{R_{a,\mathrm{Fi}}}{R_{c,\mathrm{Fb}}}\right\}_{\mathrm{SpB}} = \frac{\int_{\mathrm{SpB}} n_{\mathrm{Fi}}(\mathbf{x}) \int_E \sigma_{a,\mathrm{Fi}}(E)\phi(E,\mathbf{x})\,dE\,d\mathbf{x}}{\int_{\mathrm{SpB}} n_{\mathrm{Fb}}(\mathbf{x}) \int_E \sigma_{c,\mathrm{Fb}}(E)\phi(E,\mathbf{x})\,dE\,d\mathbf{x}} \tag{4.54}$$

represents an inverse conversion ratio of the spallation blanket and is determined by the enrichment of the fuel, $N_{\mathrm{Fi}}/N_{\mathrm{Fb}} = \varepsilon/(1-\varepsilon)$, $\varepsilon_X \leq \varepsilon \leq \varepsilon_R$, and by the fissile absorption to breeding ratio, $\sigma_{a,\mathrm{Fi}}/\sigma_{c,\mathrm{Fb}}$, as indicated by Figure 3.3 and as discussed in the context of fuel rejuvenation, Eq. (3.61).

This characterization, Eq. (4.52), may now be compared with Eq. (4.10), which applies to a system supplied only by fertile materials with no initial enrichment. As expected, Eq. (4.52) reduces to Eq. (4.10) for $\varepsilon_X = 0$ with the exception of an additional term describing the effect of fissile fuel consumption in the blanket and which had previously been neglected:

$$\{(1 - C_{\mathrm{Fi}})R_{a,\mathrm{Fi}}\}_{\mathrm{FCR}} = \left\{C_p \frac{I_p}{q_p}\left(1 - \frac{R_{a,\mathrm{Fi}}}{R_{c,\mathrm{Fb}}}\right)\right\}_{\mathrm{SpB}} \tag{4.55}$$

A system design consideration which this suggests is to carefully choose or tailor the neutron spectrum so that the $R_{a,\mathrm{Fi}}/R_{c,\mathrm{Fb}}$ term becomes as small as possible unless the pertinent power production is considered more desirable.

Examination of Eq. (4.52) shows that the appearance of the ratio $\varepsilon_R/(\varepsilon_R - \varepsilon_X)$ has the effect of reducing, say, the accelerator beam requirements, for a given companion fission reactor. For example, in the case of enriching natural uranium ($\varepsilon_X = 0.0071$) to 2% ($\varepsilon_R = 0.02$), the accelerator beam current could be reduced by about 35% compared to the supply of pure fertile materials as implied in the preceding section. On the other hand, if the depleted uranium tailings, with $\varepsilon_X = 0.002$, were to be accelerator-enriched to 2%, the beam requirements could be reduced by some 10%.

For this general case of an enriched fuel supply to the blanket, the associated power ratios may similarly be obtained directly. We use the definition of the thermal fission power, Eq. (4.15), together with the definition of the beam power, Eq. (4.22), and substitute both expressions in Eq. (4.52) to obtain

$$\frac{P_{\mathrm{fi},t}}{P_p} = \left(\frac{\varepsilon_R}{\varepsilon_R - \varepsilon_X}\right)\left(\frac{U_{\mathrm{Fi}}}{E_p}\right)\left(\frac{C_p}{1 - C_{\mathrm{Fi}}}\right)\left\{1 + \varepsilon_X \frac{R_{\mathrm{fi,Fb}}}{R_{c,\mathrm{Fb}}} - (1 - \varepsilon_X)\frac{R_{a,\mathrm{Fi}}}{R_{c,\mathrm{Fb}}}\right\}_{\mathrm{SpB}} \tag{4.56}$$

Here we again note the appearance of the enrichment ratio and additional reaction rate ratios when compared to Eq. (4.32). In the case that fission reactions in the spallation blanket occur only to a negligible extent, the thermal fission to spallation component ratio is directly obtained from Eq. (4.19), (4.22), and (4.56):

$$\frac{P_{\mathrm{fi},t}}{P_{\mathrm{sp},t}} = \left(\frac{\varepsilon_R}{\varepsilon_R - \varepsilon_X}\right)\left(\frac{U_{\mathrm{Fi}}}{U_{\mathrm{sp}}}\right)\left(\frac{C_p}{1 - C_{\mathrm{Fi}}}\right) \tag{4.57}$$

For the case of appreciable power contribution from fission reactions in the blanket, Eq. (4.39) replaces Eq. (4.19) and Eq. (4.57) would need to be replaced by the more exact formulation

$$\frac{P_{\mathrm{fi},t}}{P_{\mathrm{sp},t}} = \left(\frac{\varepsilon_R}{\varepsilon_R - \varepsilon_X}\right)\left(\frac{U_{\mathrm{Fi}}}{U_{\mathrm{sp}}}\right)\left(\frac{1}{M_{\mathrm{SpB}}}\right)\left(\frac{C_p}{1 - C_{\mathrm{Fi}}}\right)$$

$$\times\left\{1 + \varepsilon_X \frac{R_{\mathrm{fi,Fb}}}{R_{c,\mathrm{Fb}}} - (1 - \varepsilon_X)\frac{R_{a,\mathrm{Fi}}}{R_{c,\mathrm{Fb}}}\right\}_{\mathrm{SpB}} \tag{4.58}$$

where the blanket energy multiplication factor is approximately given by

$$M_{\mathrm{SpB}} \simeq \left\{1 + \frac{U_{\mathrm{Fi}}}{U_{\mathrm{Sp}}} \frac{R_{a,\mathrm{Fi}}}{I_p/q_p}\right\}_{\mathrm{SpB}} \tag{4.59}$$

Here, as previously defined, U_{Fi} includes the effect of fertile fission and fissile capture on the energy yield.

PROBLEMS

4.1. Use the several relationships of Chapter 2 to determine potentially useful analytical functional expressions for $C_{\mathrm{p}} = \mu_{\mathrm{sp}}\eta_{\mathrm{sp}}$.

4.2. Determine the trade-off between station net power and variable initial enrichment for a range of blanket multiplication.

4.3. Estimate the proton density and power density in the accelerator target.
4.4. Calculate the fission-to-beam power ratio $P_{fi,t}/P_p$ as a function of the blanket breeding capacity for various fission reactor conversion ratios (e.g., $C_{fi} = 0.5$, 0.7, 0.9).
4.5. Derive the formula for the overall system efficiency η_0, defined as the ratio between the overall net electric power output $P_{0,e}$ and all thermal power produced in the system (accelerator, target, blanket, and fission reactor).
4.6. Calculate the overall system efficiency of a spallation–fission symbiont as a function of the accelerator efficiency η_p. Assume $C_{Fi} = 0.6$ for the fission reactory, $C_p = 20$; 50; 80, $E_p = 1000$ MeV, and $G_{sp} = 2$ for the spallation component.

Chapter Five

NUCLEAR FUEL TRAJECTORIES

An expanded approach to the fuel characterization of a spallation-aided fission energy system will now be introduced. Emphasis will be placed both on the time dependence of the various fuel flow components and on the external unavailability of a part of the fuel existing in an energy system. This may consist of the *in situ* bred fuel inventory of the spallation blanket as well as the fissile fuel contained in disposed or stored spent fuel elements. In particular, we will examine the net fissile fuel flow of a spallation–fission symbiont with respect to these characteristics and assess the associated external fuel requirement and supply.

5.1. STOCKPILE INVENTORY

As a basis for the incorporation of consideration related to time dependence and fuel availability into the fuel cycle analysis of the spallation–fission symbiont, we will draw upon the concept of "equivalent fuel" and the concept of an "external stockpile." The equivalent fuel concept is a useful tool for describing fuel in terms of two types of nuclei, Fi-type fissile nuclei and Fb-type fertile nuclei, and circumvents the analytically more cumbersome analysis of itemizing the various specific nuclei involved; Appendix C provides a discussion of this concept.

The concept of an external fissile stockpile is useful in illustrating the net availability of fissile fuel associated with a synergetic system. It is most conveniently introduced by an external storage stockpile into

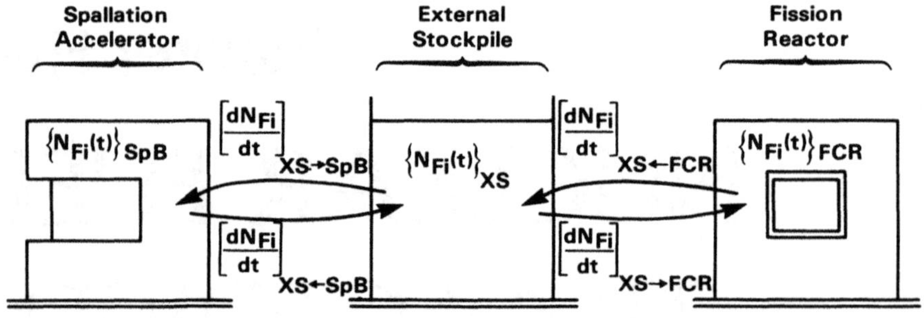

FIGURE 5.1. Instantaneous flows of fuel involving an external fuel stockpile inventory coupled to a spallation accelerator and fission converter reactor.

which all externally available bred fuel is placed and from which withdrawals are made as necessary. We illustrate such an external fissile fuel stockpile and its fuel flow connections to a spallation accelerator and companion converter fission reactor in Figure 5.1. The principal merit of the stockpile concept is that it shifts the emphasis from the amount of fuel contained in the various components of the system to the amount of fuel required from or made available to an external depository represented by the external fuel stockpile.

The deficiency/excess of fissile fuel is thus associated with the variation of the amount of fissile materials in the external stockpile as a function of time. Considerations such as the initial fuel inventory, withdrawals from the reactor core, breeding capacity of the spallation accelerator, and fissile materials contained in the spent fuel of the fission reactor can be incorporated. We consider here these factors but assume that reprocessing time intervals and other time lags are comparatively short and that losses can be neglected; this is imposed in order to make the analysis mathematically more tractable. Our emphasis will hence be on the fissile breeding and destruction processes with all transfers and processings being taken as essentially instantaneous on the time scale of the reactor operation lifetime.

The rate equation for the instantaneous change in fissile inventory in the external stockpile at any time is the starting point of our analysis. As suggested in Figure 5.1 we write this statement of mass conservation in terms of the various transfer rates:

$$\left\{\frac{d}{dt}N_{\mathrm{Fi}}(t)\right\}_{\mathrm{XS}} = \left(\left[\frac{d}{dt}N_{\mathrm{Fi}}(t)\right]_{\mathrm{XS}\leftarrow\mathrm{SpB}} - \left[\frac{d}{dt}N_{\mathrm{Fi}}(t)\right]_{\mathrm{XS}\rightarrow\mathrm{SpB}}\right) + \left(\left[\frac{d}{dt}N_{\mathrm{Fi}}(t)\right]_{\mathrm{XS}\leftarrow\mathrm{FCR}} - \left[\frac{d}{dt}N_{\mathrm{Fi}}(t)\right]_{\mathrm{XS}\rightarrow\mathrm{FCR}}\right) \tag{5.1}$$

where the subscript XS again refers to the external stockpile while SpB refers to the spallation blanket and FCR to the fission converter reactor. Since N_{Fi} designates the inventory of the external stockpile, the left-hand side of this equation describes the variation of that inventory and the right-hand side contains terms describing the fuel flow from and to the stockpile. According to the reactor representation adopted, we characterize inventory changes across a stockpile boundary by [ ] brackets whereas inventory changes within both system components and the stockpile are distinguished by { } brackets.

The stockpile is taken to possess an arbitrary initial fissile inventory, N^0_{XS}, with N^0_{FCR} and N^0_{SpB} of it removed to provide the initial fissile load for the fission reactor and the spallation blanket for start-up at $t = 0$. These two may, in principle, be fuels of significantly different enrichments with N^0_{SpB}, in particular, possibly being even fully depleted fuel. The fissile inventory in the stockpile at any point in time $t > 0$ is then given by the integration of Eq. (5.1):

$$\begin{aligned}\{N_{\mathrm{Fi}}(t)\}_{\mathrm{XS}} &= N^0_{\mathrm{XS}} - (N^0_{\mathrm{FCR}} + N^0_{\mathrm{SpB}}) \\ &+ \int_0^t \left(\left[\frac{d}{dt} N_{\mathrm{Fi}}(t)\right]_{\mathrm{XS}\leftarrow\mathrm{SpB}} - \left[\frac{d}{dt} N_{\mathrm{Fi}}(t)\right]_{\mathrm{XS}\rightarrow\mathrm{SpB}} \right) dt \\ &+ \int_0^t \left(\left[\frac{d}{dt} N_{\mathrm{Fi}}(t)\right]_{\mathrm{XS}\leftarrow\mathrm{FCR}} - \left[\frac{d}{dt} N_{\mathrm{Fi}}(t)\right]_{\mathrm{XS}\rightarrow\mathrm{FCR}} \right) dt \end{aligned} \tag{5.2}$$

This function describes the change with time of the external stockpile inventory and will be called the fissile fuel trajectory. As is evident from this equation, the important terms are the fissile flow rates which will later be related to the pertinent reaction rates in the various system components.

The fissile fuel trajectory concept could be extended to include the fertile fuel component by defining a two-component vector fuel trajectory

$$\mathbf{N}(t) = \begin{Bmatrix} N_{\mathrm{Fi}}(t) \\ N_{\mathrm{Fb}}(t) \end{Bmatrix} \tag{5.3}$$

A further refinement would be to define a trajectory for each of the actual fissile or fertile nuclei of the fuel (^{232}Th, . . . , ^{242}Pu), which would result in a system of coupled equations, necessitating the treatment by elaborate computer codes. As described in Appendix C, the equivalent fuel concept allows an intuitive and analytically more convenient method of analysis.

5.2. SIMPLIFIED TRAJECTORY

Insight from a simplified representation of the variation of $\{N_{Fi}(t)\}_{XS}$ with time—that is, the fissile fuel trajectory of the external stockpile—can be obtained by using previous results and assuming that the system is in a steady-state operational mode and that the fuel cycle is perfectly closed. Under these conditions the net fissile fuel transfer rate from the spallation blanket to the stockpile

$$\left[\frac{d}{dt}N_{Fi}(t)\right]_{XS\rightleftarrows SpB} = \left[\frac{d}{dt}N_{Fi}(t)\right]_{XS\leftarrow SpB} - \left[\frac{d}{dt}N_{Fi}(t)\right]_{XS\rightarrow SpB} \tag{5.4}$$

can be equated with the net breeding rate in the blanket by

$$\left[\frac{d}{dt}N_{Fi}(t)\right]_{XS\rightleftarrows SpB} = \left\{\frac{d}{dt}N_{Fi}(t)\right\}_{SpB} \tag{5.5}$$

Similar considerations apply to the stockpile–converter interface. Under the previously stated assumptions, the net fissile fuel transfer rate from the fission reactor to the stockpile

$$\left[\frac{d}{dt}N_{Fi}(t)\right]_{XS\rightleftarrows FCR} = \left[\frac{d}{dt}N_{Fi}(t)\right]_{XS\leftarrow FCR} - \left[\frac{d}{dt}N_{Fi}(t)\right]_{XS\rightarrow FCR} \tag{5.6}$$

is taken to be equal to the net fissile fuel consumption rate in the fission reactor, Eqs. (4.2) and (4.5)

$$\left[\frac{d}{dt}N_{Fi}(t)\right]_{XS\rightleftarrows FCR} = \left\{\frac{d}{dt}N_{Fi}(t)\right\}_{FCR} \tag{5.7}$$

Note that the double arrows refer to the net fuel flow.

For the spallation component, we find by extension of Eq. (4.6) from Eqs. (5.4) and (5.5)

$$\left[\frac{d}{dt}N_{Fi}(t)\right]_{XS\rightleftarrows SpB} = \{C_pI_p/q_p\}_{SpB} \tag{5.8}$$

Similarly, we obtain for the fission component from Eqs. (5.6) and (5.7) with Eqs. (4.2)–(4.5)

$$\left[\frac{d}{dt}N_{\mathrm{Fi}}(t)\right]_{\mathrm{XS}\rightleftarrows\mathrm{FCR}} = \{R_{c,\mathrm{Fb}} - R_{a,\mathrm{Fi}}\}_{\mathrm{FCR}}$$

$$= \{(C_{\mathrm{Fi}} - 1)R_{a,\mathrm{Fi}}\}_{\mathrm{FCR}} \tag{5.9}$$

The stockpile inventory change, Eq. (5.1), is therefore written as

$$\left\{\frac{d}{dt}N_{\mathrm{Fi}}(t)\right\}_{\mathrm{XS}} = \left\{C_p\frac{I_p}{q_p}\right\}_{\mathrm{SpB}} - \{(1 - C_{\mathrm{Fi}})R_{a,\mathrm{Fi}}\}_{\mathrm{FCR}} \tag{5.10}$$

and the fissile fuel trajectory is obtained by integration with respect to time, Eq. (5.2):

$$\{N_{\mathrm{Fi}}(t)\}_{\mathrm{XS}} = N^0_{\mathrm{XS}} - (N^0_{\mathrm{FCR}} + N^0_{\mathrm{SpB}})$$

$$+ \int_0^t \{C_pP_p/E_p\}_{\mathrm{SpB}}\, \mathrm{dt} - \int_0^t \{(1 - C_{\mathrm{Fi}})P_{\mathrm{fi},t}/U_{\mathrm{Fi}}\}_{\mathrm{FCR}}\, dt \tag{5.11}$$

Here we have incorporated the accelerator proton beam power P_p and the fission reactor thermal power $P_{\mathrm{fi},t}$ defined previously in Eqs. (4.15) and (4.22).

In general, all terms appearing in the integrand have a time dependence attributable to two distinct origins. One type is associated with the long-term change in the isotopic composition of the spallation blanket and fission core. The other is related to the operation of the system; clearly, P_p and $P_{\mathrm{fi},t}$ are zero if the respective system component is shut down for repair, fuel management operation, or inspection, but attain their nameplate power output or fractions thereof during operation.

In our analysis here we account for the first type of time dependence by the use of appropriately averaged values giving the fissile trajectory of the external stockpile as

$$\{N_{\mathrm{Fi}}(t)\}_{\mathrm{XS}} = N^0_{\mathrm{XS}} - (N^0_{\mathrm{SpB}} + N^0_{\mathrm{FCR}})$$

$$+ \frac{C_p}{E_p}\int_0^t P_p(t)\, dt - \frac{1 - C_{\mathrm{Fi}}}{U_{\mathrm{Fi}}}\int_0^t P_{\mathrm{fi},t}(t)\, dt \tag{5.12}$$

The second type of time dependence will be accounted for by the introduction of load factors L_j for each component according to

$$L_{\mathrm{SpB}} = \frac{\int_{\Delta t} P_p(t)\, dt}{P_p \Delta t} \leq 1 \tag{5.13}$$

and

$$L_{\mathrm{FCR}} = \frac{\int_{\Delta t} P_{\mathrm{fi},t}(t)\, dt}{P_{\mathrm{fi},t} \Delta t} \leq 1 \tag{5.14}$$

where $P_p(t)$ and $P_{\mathrm{fi},t}(t)$ are the operational power levels as a function of time within a given time interval Δt, usually a year, and P_p and $P_{\mathrm{fi},t}$ are the nameplate values of the accelerator beam power and the thermal fission reactor power, respectively. Normally L_{SpB} and L_{FCR} are slowly varying functions of time on the time scale of system lifetime but here they will be taken as constant. This then yields the fissile fuel trajectory equation for the spallation–fission symbiont:

$$\{N_{\mathrm{Fi}}(t)\}_{\mathrm{XS}} = N^0_{\mathrm{XS}} - (N^0_{\mathrm{SpB}} + N^0_{\mathrm{FCR}}) + \left[\left\{\frac{C_p}{E_p} P_p\right\}_{\mathrm{SpB}} L_{\mathrm{SpB}} - \left\{\frac{1 - C_{\mathrm{Fi}}}{U_{\mathrm{Fi}}} P_{\mathrm{fi},t}\right\}_{\mathrm{FCR}} L_{\mathrm{FCR}}\right] t \tag{5.15}$$

According to this introductory analysis, the fissile inventory—for this equilibrium fuel cycle case—increases or decreases linearly with time depending upon whether the accelerator breeds fissile material faster or slower than the companion fission reactor consumes it. We illustrate this in Figure 5.2.

The special case of fuel self-sufficiency of the spallation symbiont examined here provides an interesting reference point. Fuel self-sufficiency, would according to Eq. (5.15) imply

$$\left\{\frac{C_p}{E_p} P_p L\right\}_{\mathrm{SpB}} = \left\{\frac{1 - C_{\mathrm{Fi}}}{U_{\mathrm{Fi}}} \mathrm{P}_{\mathrm{fi},t} L\right\}_{\mathrm{FCR}} \tag{5.16}$$

This expression resembles that of Eq. (4.31) applicable for the case of $L_{\mathrm{SpB}} = L_{\mathrm{FCR}} = 1$. Nevertheless, the analysis in this section casts the analysis of Chapter 4 into a new representation, the merits of which will become evident as we remove in the following some of the imposed assumptions

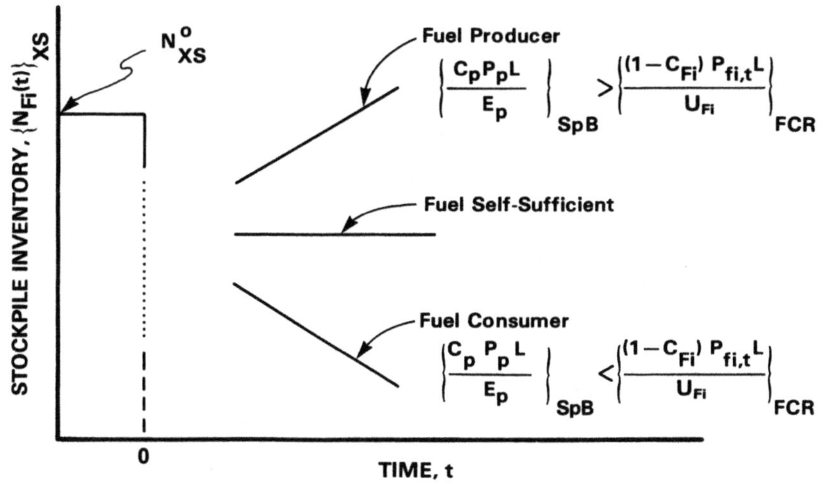

FIGURE 5.2. Generalized fuel trajectories of a spallation–fission symbiont for sufficient time after startup.

5.3. DISCONTINUOUS TRAJECTORIES

One of the implicit assumptions on which all previous flow considerations have been based is that of a continuous fuel flow between the various system components. While this may be adequate for some blanket and reactor concepts such as the molten salt reactors and those based on on-line refueling, it is unsatisfactory in many other cases. We consider this alternative representation next.

In Eq. (5.8) we have equated the net rate at which fissile fuel is supplied from the spallation blanket to the stockpile with the net rate at which fuel is produced in the blanket. The pertinent contribution of the spallation component to the stockpile trajectory is shown in Figure 5.3a by the dashed line as obtained by taking the spallation component of Eq. (5.15):

$$[N_{\mathrm{Fi}}(t)]_{\mathrm{XS}\rightleftarrows\mathrm{SpB}} = -N^0_{\mathrm{SpB}} + \left\{\frac{C_p}{E_p} P_p L\right\}_{\mathrm{SpB}} t \tag{5.17}$$

Simultaneously, the fissile inventory of the blanket remains constant (dashed line in Figure 5.3b):

$$\{N_{\mathrm{Fi}}(t)\}_{\mathrm{SpB}} = N^0_{\mathrm{SpB}} \tag{5.18}$$

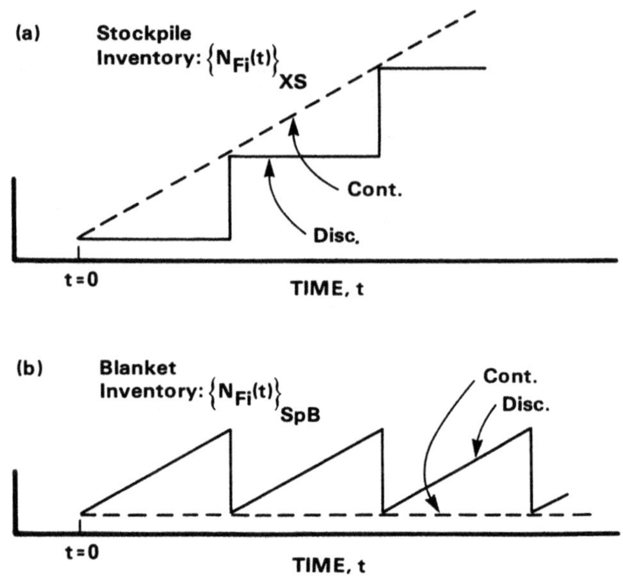

FIGURE 5.3. Illustration comparing continuous (Cont.) and discrete (Disc.) fissile inventories for the stockpile and for the accelerator blanket.

If, alternatively, the bred fuel is allowed to accumulate in the spallation blanket during a time interval τ_a and then is instantly discharged, then the amount of externally available fuel increases stepwise (solid line Figure 5.3a), as described by

$$[N_{\mathrm{Fi}}(t)]_{\mathrm{XS}\rightleftarrows\mathrm{SpB}} = -N^0_{\mathrm{SpB}} + \left\{\frac{C_p}{E_p} P_p L m \tau_a\right\}_{\mathrm{SpB}} \tag{5.19}$$

with time t contained in the interval $m\tau_a \leq t < (m+1)\tau_a$, $m = 0, 1, 2, \ldots$.

The difference between the amount of externally available fissile fuel in the continuous case, and in the discontinuous case, leads to a difference in the time variation of the blanket inventory which now is represented by a saw-tooth shaped curve of Figure 5.3b, when compared to the previously constant inventory, Eq. (5.18):

$$\{N_{\mathrm{Fi}}(t)\}_{\mathrm{SpB}} = N^0_{\mathrm{SpB}} + \left\{\frac{C_p}{E_p} P_p L (t - m\tau_a)\right\}_{\mathrm{SpB}} \tag{5.20}$$

The same considerations apply to the fission reactor if its makeup fuel is periodically loaded in discrete steps. For a refueling interval of τ_r

the fissile inventory in the fission reactor will vary according to

$$\{N_{\text{Fi}}(t)\}_{\text{FCR}} = N^0_{\text{FCR}} - \left\{\frac{1 - C_{\text{Fi}}}{U_{\text{Fi}}} P_{\text{fi},t} L \cdot (t - m\tau_r)\right\}_{\text{FCR}} \tag{5.21}$$

and the external stockpile will consequently be depleted according to

$$[N_{\text{Fi}}(t)]_{\text{XS}\rightleftarrows\text{FCR}} = -N^0_{\text{FCR}} - \left\{\frac{1 - C_{\text{Fi}}}{U_{\text{Fi}}} P_{\text{fi},t} L \cdot (m\tau_r)\right\}_{\text{FCR}} \tag{5.22}$$

It is convenient, though somewhat artificial, to take the accumulation interval τ_a and the refueling interval τ_r associated with the spallation blanket and the converter reactor, respectively, as the same, $\tau_a = \tau_r = \tau$. Then the fissile fuel trajectory

$$\{N_{\text{Fi}}(t)\}_{\text{XS}} = N^0_{\text{XS}} + [N_{\text{Fi}}(t)]_{\text{XS}\rightleftarrows\text{SpB}} + [N_{\text{Fi}}(t)]_{\text{XS}\rightleftarrows\text{FCR}} \tag{5.23}$$

takes the simple form

$$\{N_{\text{Fi}}(t)\}_{\text{XS}} = N^0_{\text{XS}} - (N^0_{\text{SpB}} + N^0_{\text{FCR}}) + \left(\left\{\frac{C_p}{E_p} P_p L\right\}_{\text{SpB}} - \left\{\frac{1 - C_{\text{Fi}}}{U_{\text{Fi}}} P_{\text{fi},t} L\right\}_{\text{FCR}}\right) m\tau \tag{5.24}$$

Evidently Eq. (5.15) constitutes an envelope of step functions of varying time intervals τ with the upper envelope applicable for a net fuel-producing symbiont and the lower envelope for a net fuel-consuming symbiont. For the case of a fuel self-sufficient symbiont, Eqs. (5.15) and (5.24) do coincide. These results would be modified if $\tau_r \neq \tau_a$. The point to note, however, is that the fuel management procedure may introduce transient effects on the external fuel availability and requirement which are not encountered in a continuous representation.

5.4. TRAJECTORY DYNAMICS

In the previous section we have considered discontinuous and continuous fuel management schemes as they relate to the external fuel

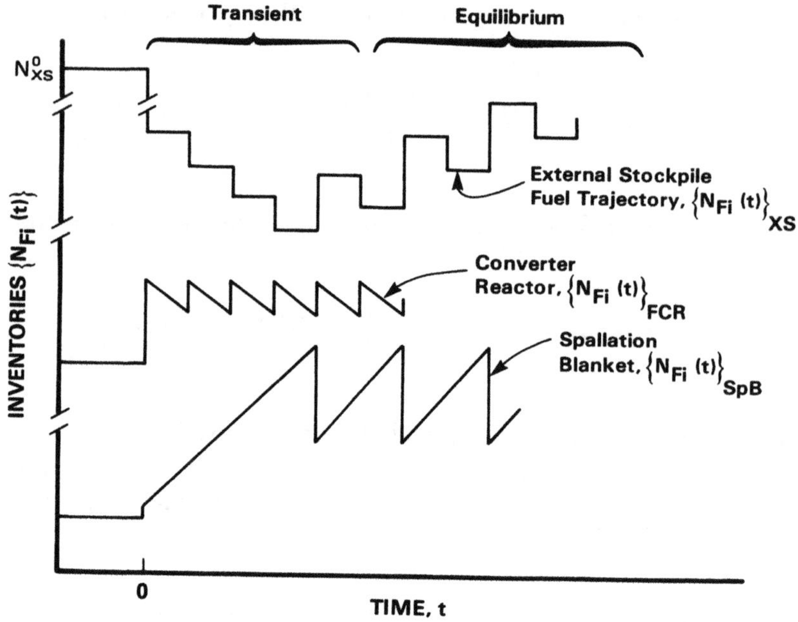

FIGURE 5.4. Illustration of discontinuous trajectories which lead to a dominant nonlinear stockpile inventory trajectory.

requirement/supply trajectory. This has been done under the implicit assumption that both the fuel management scheme and the component parameter characterizations are constant with time over the system lifetime. We now examine again the trajectory representation and include the effect of preequilibrium conditions which appear between the time of reactor start-up and the time the system reaches its operational fuel-cycle equilibrium.

As an illustrative example, we consider a spallation–fission symbiont which breeds all or part of its own inventory requirement *in situ*. While the component characterization is again taken to be constant, the effect of an *in situ* blanket inventory build-up on the external fuel stockpile inventory is examined. The result is graphically suggested in Figure 5.4 for discontinuous fuel flow. The analytical procedure for the continuous case is indicated in the following.

For a spallation breeding blanket from which bred fuel is continuously extracted, it is generally satisfactory to assume that at any time the amount of fuel removed is proportional to the blanket inventory at that time; that is

$$\left[\frac{d}{dt}N_{Fi}(t)\right]_{XS\leftarrow SpB} = \frac{1}{\tau_s}\{N_{Fi}(t)\}_{SpB} \qquad (5.25a)$$

and further assume

$$\left[\frac{d}{dt} N_{\mathrm{Fi}}(t)\right]_{\mathrm{XS}\rightarrow\mathrm{SpB}} = 0 \tag{5.25b}$$

where τ_s is a characteristic time constant representing the average residence time of fissile fuel in the spallation blanket. The blanket inventory variation is therefore given by the balance relation

$$\left\{\frac{d}{dt} N_{\mathrm{Fi}}(t)\right\}_{\mathrm{SpB}} = \left\{\frac{C_p}{E_p} P_p(t)\right\}_{\mathrm{SpB}} - \frac{1}{\tau_s}\{N_{\mathrm{Fi}}(t)\}_{\mathrm{SpB}} \tag{5.26}$$

where the first term is the net production rate and second term the net removal rate. This equation may be solved for an initial condition which we take as the fissile inventory at $t = 0$ as

$$\{N_{\mathrm{Fi}}(0)\}_{\mathrm{SpB}} = N^0_{\mathrm{SpB}} \tag{5.27}$$

giving the time-dependent blanket inventory $\{N_{\mathrm{Fi}}(t)\}_{\mathrm{SpB}}$ as

$$\{N_{\mathrm{Fi}}(t)\}_{\mathrm{SpB}} = N^0_{\mathrm{SpB}} \exp(-t/\tau_s) + \left\{\tau_s \frac{C_p}{E_p} P_p L\right\}_{\mathrm{SpB}} [1 - \exp(-t/\tau_s)] \tag{5.28}$$

The eventual equilibrium blanket fissile inventory N^∞_{SpB} for $t \gg \tau_s$ is thus given by

$$N^\infty_{\mathrm{SpB}} = \left\{\tau_s \frac{C_p}{E_p} P_p L\right\}_{\mathrm{SpB}} \tag{5.29}$$

and is independent of the initial inventory.

With the above results, we may now return to the determination of the stockpile trajectory $\{N_{\mathrm{Fi}}(t)\}_{\mathrm{XS}}$, Eq. (5.2). For the case of a perfectly closed spallation–stockpile fuel flow interface, the first integrand of Eq. (5.2) is represented by the net fissile flow rate expression derived from Eqs. (5.25) and (5.26)

$$\left[\frac{d}{dt} N_{\mathrm{Fi}}(t)\right]_{\mathrm{XS}\rightleftarrows\mathrm{SpB}} = \frac{1}{\tau_s} \mathrm{N}^0_{\mathrm{SpB}} \exp(-t/\tau_s) + \left\{\frac{C_p}{E_p} P_p L\right\}_{\mathrm{SpB}} \cdot [1 - \exp(-t/\tau_s)] \tag{5.30}$$

The fissile flow for the reactor–stockpile interface is again taken to be represented by Eqs. (5.7) and (5.9). Hence, substituting Eqs. (5.9) and (5.30) in Eq. (5.2) then yields

$$
\begin{aligned}
\{N_{\mathrm{Fi}}(t)\}_{\mathrm{XS}} = N_{\mathrm{XS}}^{0} - (N_{\mathrm{FCR}}^{0} + N_{\mathrm{SpB}}^{0}) \\
+ \int_0^t \left[\left\{ \frac{1}{\tau_s} N_{\mathrm{SpB}}^{0} \exp(-t/\tau_s) \right\}_{\mathrm{SpB}} \right. \\
\left. + \left\{ \frac{C_p}{E_p} P_p L [1 - \exp(-t/\tau_s)] \right\}_{\mathrm{SpB}} \right] dt \\
+ \int_0^t \left\{ \frac{(C_{\mathrm{Fi}} - 1)}{U_{\mathrm{Fi}}} P_{\mathrm{Fi},t} L \right\}_{\mathrm{FCR}} dt
\end{aligned}
\tag{5.31}
$$

as a more complete representation by accounting for the inventory buildup in the spallation blanket.

5.5. GENERAL TRAJECTORY CHARACTERIZATION

The more comprehensive variation in the external stockpile follows from the integration of Eq. (5.31) yielding specifically the trajectory for the case that part of the inventory is bred *in situ*. The general result is

$$
\{N_{\mathrm{Fi}}(t)\}_{\mathrm{XS}} = f_0 + f_1 t - f_2 [1 - \exp(-t/\tau_s)]
\tag{5.32}
$$

where the coefficient functions f_i are defined by the following

$$
f_0 = N_{\mathrm{XS}}^{0} - (N_{\mathrm{SpB}}^{0} + N_{\mathrm{FCR}}^{0})
\tag{5.33a}
$$

$$
f_1 = \left\{ \frac{C_p}{E_p} P_p L \right\}_{\mathrm{SpB}} - \left\{ \frac{1 - C_{\mathrm{Fi}}}{U_{\mathrm{Fi}}} P_{\mathrm{fi},t} L \right\}_{\mathrm{FCR}}
\tag{5.33b}
$$

$$
f_2 = \left\{ \tau_s \frac{C_p}{E_p} P_p L \right\}_{\mathrm{SpB}} - N_{\mathrm{SpB}}^{0} = N_{\mathrm{SpB}}^{\infty} - N_{\mathrm{SpB}}^{0}
\tag{5.33c}
$$

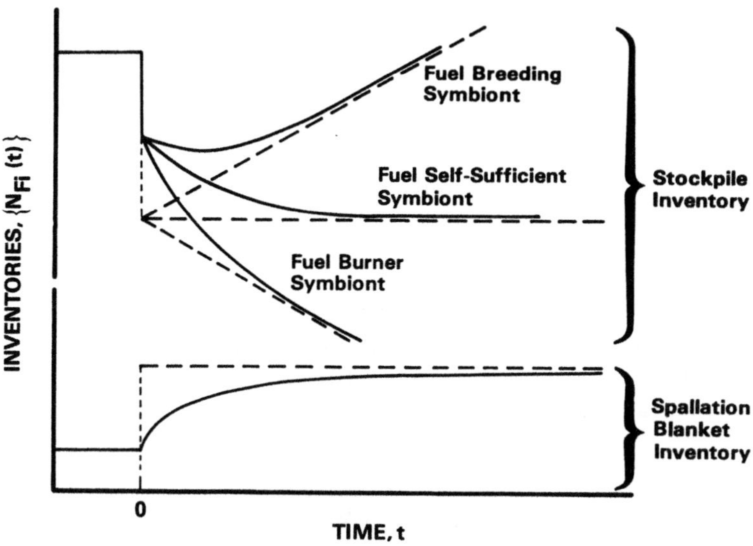

FIGURE 5.5. Schematic illustration of classes of stockpile fissile inventory trajectories.

These three system-specific functions characterize particular features of the spallation–fission symbiont of general applicability. The first function f_0 represents the stockpile inventory immediately after start-up. The second function f_1, can be identified with the asymptotic behavior of the net fissile fuel transfer rate from the symbiont components to the stockpile which is reached once all system parameters have attained their equilibrium values, therefore f_1 characterizes the system with respect to its fuel requirement/supply categorization:

$$f_1 = \begin{cases} > 0, & \text{system is a net fuel supplier} \\ = 0, & \text{system is asymptotically fuel self-sufficient} \\ < 0, & \text{system is a net fuel consumer} \end{cases} \qquad (5.34)$$

The third term describes the effect of the *in situ* inventory built up here in the spallation blanket from $N^0_{\mathrm{Fi,SpB}}$ to the equilibrium inventory $N^\infty_{\mathrm{Fi,SpB}}$. Evidently, the preequilibrium features pertinent to the blanket inventory buildup can be eliminated either by instantaneous bred fuel removal, $\tau_s = 0$, or by the initial supply of the equilibrium inventory, $N^0_{\mathrm{SpB}} = N^\infty_{\mathrm{SpB}}$.

Figure 5.5 displays typical symbiont fuel trajectories. This figure is to be compared with Figure 5.2 illustrating the effect of the preequilibrium features associated with the inventory buildup in the spallation blanket. In both figures, the asymptotes have the same slope but are displaced by $(N^{\infty}_{\mathrm{SpB}} - N^{0}_{\mathrm{SpB}})$:

$$\lim_{t\to\infty} \{N_{\mathrm{Fi}}(t)\}_{\mathrm{XS}} = (f_0 - f_2) + f_1 t \tag{5.35}$$

with f_1 describing the asymptotic net fuel transfer rate to the stockpile and the difference $(f_0 - f_2)$ representing the initial inventory reduction reflecting the asymptotic spallation blanket inventory and the initial fission converter fuel loading:

$$f_0 - f_2 = N^{0}_{\mathrm{XS}} - (N^{\infty}_{\mathrm{SpB}} + N^{0}_{\mathrm{FCR}}) \tag{5.36}$$

This suggests the important conclusion that the effect of any component inventory on the external fuel requirement is, in the long term, independent of whether the inventory is externally supplied or bred *in situ*. This is illustrated by the dashed lines in Figure 5.5.

Furthermore, the upper curve in Figure 5.5, labeled fuel breeding symbiont, is to be compared with Figure 5.4. Both cases are equivalent except for the fuel management scheme, which is discontinuous in Figure 5.4 and continuous in Figure 5.5. The histogram in Figure 5.4 and the upper curve in Figure 5.5 show a characteristic minimum in the trajectory which reflects the preequilibrium blanket inventory built up. To emphasize the similarity we have also plotted in Figure 5.5 the blanket inventory trajectory for continuous fuel management which compares to that in Figure 5.4 for the discontinuous case. These similarities suggest that the continuous description may be regarded as a good approximation to the discontinuous fuel flow since the important features of the various trajectories are retained.

5.6. TRAJECTORY ITEMIZATION

So far we have assumed that every fissile nucleus supplied to the fission reactor is eventually consumed by neutron absorption. This defines a completely closed, loss-free fuel cycle in which all nuclear fuel contained in spent fission reactor fuel elements is recovered and fully recycled. This is described by the net fuel requirement expression

$$\left[\frac{d}{dt}N_{\mathrm{Fi}}(t)\right]_{\mathrm{XS}\rightleftarrows\mathrm{FCR}} = \{R_{c,\mathrm{Fb}} - R_{a,\mathrm{Fi}}\}_{\mathrm{FCR}}$$

$$= \left\{\frac{C_{\mathrm{Fi}} - 1}{U_{\mathrm{Fi}}} P_{\mathrm{fi},t}(t)\right\}_{\mathrm{FCR}} \tag{5.37}$$

The pertinent contribution of the fission reactor to the stockpile trajectory is therefore

$$[N_{\mathrm{Fi}}(t)]_{\mathrm{XS}\rightleftarrows\mathrm{FCR}} = -N^0_{\mathrm{FCR}} - \left\{\frac{1 - C_{\mathrm{Fi}}}{U_{\mathrm{Fi}}} P_{\mathrm{fi},t} L\right\}_{\mathrm{FCR}} t \tag{5.38}$$

In order to extend the fuel flow analysis to other than fully closed fuel cycles, it is necessary to itemize each individual component of the actual fuel flow. The net amount of loaded fuel then appears as the difference between the fissile fuel loaded in fresh fuel elements and that contained in spent fuel elements discharged from the core. With τ_s as the average residence time of fuel elements in the core, the in/out components become for the continuous transfer formulation

$$\left[\frac{d}{dt}N_{\mathrm{Fi}}(t)\right]_{\mathrm{XS}\rightarrow\mathrm{FCR}} = \left\{\frac{N^0}{\tau_s}\right\}_{\mathrm{FCR}} \tag{5.39}$$

and

$$\left[\frac{d}{dt}N_{\mathrm{Fi}}(t)\right]_{\mathrm{XS}\leftarrow\mathrm{FCR}} = \left\{\frac{N^0}{\tau_s} - \frac{1 - C_{\mathrm{Fi}}}{U_{\mathrm{Fi}}} P_{\mathrm{fi},t} L\right\}_{\mathrm{FCR}} \tag{5.40}$$

A point to observe here is that the fissile fuel inventory of fresh fuel, $\{N^0/\tau_s\}$, is actually not constant over the fission reactor lifetime: the initial isotopic content of fissile nuclei has to be increased as the reactor approaches its equilibrium operational characteristics in order to compensate for the increasing inventory of neutron-absorbing fission products and their decay products. A more exact fuel analysis would have to account for this increase of the initial fissile inventory with time.

5.7. CYCLE DEPENDENCIES

With the previous itemization, Eqs. (5.39) and (5.40), of the converte trajectory component for a fully closed fuel cycle, we now consider othe fuel management schemes. For this purpose we modify Eq. (5.40) t incorporate the back-end options for the fuel cycle as follows:

$$\left[\frac{d}{dt}N_{\mathrm{Fi}}(t)\right]_{\mathrm{XS}\leftarrow\mathrm{FCR}} = \begin{cases} 0, & \text{once-through cycle} \\ \left\{\dfrac{N^0}{\tau_s} - \dfrac{P_{\mathrm{fi},t}}{U_{\mathrm{Fi}}}L\right\}, & U \text{ recycle} \\ \left\{\dfrac{C_{\mathrm{Fi}}}{U_{\mathrm{Fi}}}P_{\mathrm{fi},t}L\right\}_{\mathrm{FCR}}, & \text{Pu recycle} \\ \vdots & \\ \text{etc.} & \end{cases} \tag{5.41}$$

Here the once-through cycle refers to the management scheme in whicl spent fuel rods are stored and fissile fuel is not recycled. The second cas illustrates fuel management in which the fissile fuel in the fresh fue elements is uranium (^{233}U, ^{235}U) and the fertile fuel is ^{238}U with the bre plutonium not recycled. The third case is the corresponding one in whicl only Pu is recycled. Losses occurring in the course of the fuel recyclin may be accounted for by a recycling factor ε_c. For the fully closed fue cycle we write

$$\left[\frac{d}{dt}N_{\mathrm{Fi}}(t)\right]_{\mathrm{XS}\leftarrow\mathrm{FCR}} = \varepsilon_c\left\{\frac{N^0}{\tau_s} - \frac{1-C_{\mathrm{Fi}}}{U_{\mathrm{Fi}}}\mathrm{P}_{\mathrm{fi},t}L\right\}_{\mathrm{FCR}} \tag{5.42}$$

and similarly for other fuel management schemes. Evidently, $\varepsilon_c =$ describes the loss-free case while $\varepsilon_c = 0$ describes the once-through case

Each option for the back end of the fuel cycle can be incorporate as a separate formulation of the fissile fuel trajectory component associate with the converter reactor. In order to avoid this additional complexit of the analysis, we will continue to use the previously developed formula tions for the loss-free, completely closed cycle by generalizing the notio of the conversion ratio. For this purpose we define a fuel cycle conversio ratio C_c by

$$\left\{\frac{C_c - 1}{U_{\text{fi}}} P_{\text{fi},t}(t)\right\}_{\text{FCR}} = -\frac{N^0_{\text{FCR}}}{\tau_s} + \left[\frac{d}{dt} N_{\text{Fi}}(t)\right]_{\text{XS}\leftarrow\text{FCR}} \tag{5.43}$$

where the first term on the right-hand side describes the fuel supply rate to the reactor and the second the fuel-management-dependent recycling rate as suggested by Eqs. (5.41) and (5.42). Clearly $C_c = C_{\text{Fi}}$ for the ideal loss-free complete-recycle case. The general expression for the fuel cycle conversion ratio follows directly from Eq. (5.43):

$$C_c = 1 - \left\{\frac{U_{\text{Fi}}}{P_{\text{fi},t} L}\right\}_{\text{FCR}} \left(\frac{N^0_{\text{FCR}}}{\tau_s} - \left[\frac{d}{dt} N_{\text{Fi}}(t)\right]_{\text{XS}\leftarrow\text{FCR}}\right) \tag{5.44}$$

where the appropriate expressions for the recycle term may be introduced for each case of interest. With C_c thus defined, all further analyses could be performed as if the converter reactor was associated with a fully closed fuel cycle with the pertinent fuel losses accounted for by the use of C_c; hence, C_c assumes the role of C_{Fi} in the conventional rate expression.

As a further comment we point out that the itemization of in/out fuel flows can be used to incorporate hold-up times such as the average time interval τ_c which elapses between fuel discharge from the converter reactor and its reappearance in a fresh fuel element. This would be accomplished by using

$$\left[\frac{d}{dt} N_{Fi}(t)\right]_{\text{XS}\rightleftarrows\text{FCR}} = -\left[\frac{d}{dt} N_{\text{Fi}}(t)\right]_{\text{XS}\rightarrow\text{FCR}} + \left[\frac{d}{dt} N_{\text{Fi}}(t - \tau_c)\right]_{\text{XS}\leftarrow\text{FCR}} \tag{5.45}$$

where the first term on the right-hand side describes the rate at which fissile fuel is charged into the fission reactor and the second term is the rate at which fissile fuel contained in fuel elements discharged at time $(t - \tau_c)$ becomes available for reuse at time t. Clearly this time lag will have a major impact on the trajectory since the delayed recycle of fissile fuel is associated with a fuel cycle inventory of the order of τ_c times the annual discharge rate. This is another internally built up inventory and has the same effect on the trajectory as the *in situ* bred inventory of the breeder blanket. For the case of a fuel-breeding symbiont depicted in Figure 5.5 the trajectory associated with a finite lag time in the recycling is illustrated in Figure 5.6.

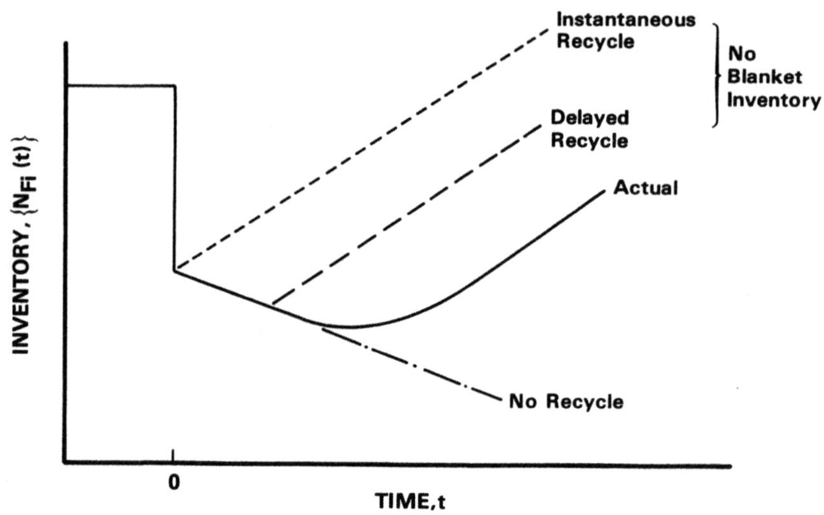

FIGURE 5.6. Schematic fuel trajectories illustrating selected fuel cycle characteristics.

5.8. TRAJECTORY PARAMETRIZATION

The stockpile fissile fuel trajectory can be used to identify selected criteria to be used in system design and planning. For example, consider a net fissile fuel-producing symbiont, Figure 5.6, consisting of a spallation breeder and one companion fission converter reactor. Then, an interesting consideration is the time required to supply the initial inventory needed for start-up of a second converter reactor, and a third, etc. until the support capacity of the accelerator has been exhausted. The fuel trajectory of interest is given by

$$\{N_{\mathrm{Fi}}(t)\}_{\mathrm{XS}} = \{N_{\mathrm{Fi}}(t)\}_{\mathrm{XS,SpB}} + \{N_{\mathrm{Fi}}(t)\}_{\mathrm{XS},0} + \{N_{\mathrm{Fi}}(t - t_1)\}_{\mathrm{XS},1} + \{N_{\mathrm{Fi}}(t - t_2)\}_{\mathrm{XS},2} + \cdots \tag{5.46}$$

where the subscripts designate the components from the spallation breeder (SpB) and the companion fission converter reactor (0), from the first additional reactor (1) introduced at time t_1, etc. Clearly,

$$\{N_{\mathrm{Fi}}(t)\}_{\mathrm{XS}} \geq 0 \tag{5.47}$$

limits the number of fission reactors which can be supported and hence the rate of their introduction.

From the trajectory of the symbiont under consideration, one can calculate the time required for the external stockpile to reach its initial fissile inventory level. This feature would, in effect, characterize the case of a symbiont that has produced a sufficient fissile surplus to supply a new identical symbiont system with the initial start-up inventory requirement. Defining this doubling time interval by τ_2 with $t = 0$ marking the system start-up, we need to solve the appropriate stockpile trajectory equation for the condition

$$\{N_{\mathrm{Fi}}(\tau_2)\}_{\mathrm{XS}} = N^0_{\mathrm{XS}} \tag{5.48}$$

As described in Section 5.5, the trajectory can be represented in the form

$$\{N_{\mathrm{Fi}}(t)\}_{\mathrm{XS}} = f_0 + f_1 t - f_2 g(t) \tag{5.49}$$

where f_0 accounts for the initial inventories, while the linear term describes the asymptotic system behavior; the third term accounts for the transient system features and may display a complex behavior depending on the preequilibrium time dependence of the system and operational parameters but possessing the asymptotic limit

$$\lim_{t \to \infty} g(t) = 1 \tag{5.50}$$

For the case that the transient feature is solely due to the spallation blanket inventory buildup we have

$$g(t) = 1 - \exp(-t/\tau_s) \tag{5.51}$$

and for this case, the time interval τ_2 is obtained from

$$(f_0 - N^0_{\mathrm{XS}}) + f_1\tau_2 - f_2[1 - \exp(-\tau_2/\tau_s)] = 0 \tag{5.52}$$

This transcendental equation could be solved by iteration unless there is reason to believe that

$$\exp(-\tau_2/\tau_s) \ll 1 \tag{5.53}$$

in which case we obtain directly

$$\tau_2 = \frac{N^0_{XS} + f_2 - f_0}{f_1}$$

$$= \frac{N^0_{FCR} + N^\infty_{SpB}}{f_1} \tag{5.54}$$

with the asymptotic transfer rate f_1 used in Eq. (5.33b). Note that this τ_2 may be viewed as an initial fissile fuel doubling time of the symbiont since at this point in time the external stockpile contains the same amount of fuel it did at start-up.

Another useful characteristic is the time interval for the trajectory of Figure 5.6 to reach its minimum. This time interval is important because it identifies the total initial fissile supply required to start and operate the symbiont until it begins to make a positive fissile return contribution to the stockpile. The time τ_m when the minimum occurs for the symbiont characterized by Eqs. (5.49) and (5.51) can—upon differentiation—be shown to be given by

$$\tau_m = \tau_s \ln\left(\frac{f_2}{f_1\tau_s}\right) \tag{5.55}$$

$$= \tau_s \ln \frac{\left\{\dfrac{C_p}{E_p} P_p L\right\}_{SpB} + \left\{\dfrac{1 - C_{Fi}}{U_{Fi}} P_{fi,t} L\right\}_{FCR}}{N^\infty_{SpB} - N^0_{SpB})\tau_s}$$

Inserting this expression of time into Eq. (5.49) then yields the minimum inventory requirement as

$$\{N_{Fi}\}_{XS,m} = N^0_{XS} - \{N_{Fi}(\tau_m)\}_{XS} \tag{5.56}$$

There is another useful feature associated with the fuel trajectory concept. If the initial inventory of the external stockpile, N^0_{XS}, is identified with the fissile fuel available to a system, then the fuel trajectory concept can be used to optimize a nuclear energy deployment strategy such that, by appropriate combination of spallation and other breeders with fission reactors, the desired expansion rate can be met with the available resources. The overall trajectory then is

$$\{N_{\mathrm{Fi}}(t)\}_{\mathrm{XS}} = \sum_j \{N_{\mathrm{Fi}}(t - t_j)\}_{\mathrm{XS},j} + N_{\mathrm{ext}}(t) \tag{5.57}$$

where t_j is the point in time when the jth component of the system, say a spallation breeder or a converter reactor, is introduced and where $N_{\mathrm{ext}}(t)$ describes the possible external fuel supply. Clearly the condition to be imposed is

$$\{N_{\mathrm{Fi}}(t)\}_{\mathrm{XS}} \geq 0 \quad \text{for all } t > t_0 \tag{5.58}$$

with the net system power output, or any other system feature of interest being the quantity to be optimized. This illustrates how the fuel trajectory concept might be used for various energy planning strategies.

PROBLEMS

5.1. Determine the external stockpile fissile inventory in which the load factor varies smoothly from 0.40 to an asymptotic value of 0.85 at an arbitrary rate.

5.2. Numerically evaluate the stockpile trajectory for a "realistic" accelerator/reactor system which illustrates the three distinct trajectory components of Figure 5.5.

5.3. Extend the trajectory analysis by considering fissile fuel removal from the fission reactor at a rate proportional to its inventory as associated with fission breeder reactors.

5.4. Consider a spallation–fission symbiont discussed in Section 4.6. Calculate the net fissile fuel transfer rate between the fission reactor and the stockpile

$$\left[\frac{d}{dt}N_{\mathrm{Fi}}(t)\right]\mathrm{XS} \rightleftarrows \mathrm{FCR[kg/y]}$$

and determine therefrom the parameters for the spallation component required to make the system fuel self-sufficient. Take standard values for the parameter and plot the result in terms of C_p vs P_p.

5.5. Extend the fissile fuel flow description, Eq. (5.40) and Eq. (5.41), to the case of a discontinuous fuel management scheme with τ_r being the time interval between two refueling procedures.

5.6. Consider a 3000-MW fission reactor with standard parameters ($C_{\mathrm{Fi}} = 0.5$, $L_{\mathrm{FCR}} = 0.7$, $U_{\mathrm{Fi}} = 190$ MeV, $\tau_s = 3$ years, $N^0_{\mathrm{FCR}} = 2400$ kg). Calculate (i) the fuel load rate into the core, (ii) the gross and the net fissile fuel consumption rate in the core. Calculate the cycle conversion ratio (i) for 10% reprocessing loss and (ii) for the once-through management and discuss the results.

Part III

D–T FUSION–FISSION INTEGRATION

Chapter Six

THE (D–T) FUSION–FISSION SYMBIONT

One of the synergetic systems concepts referred to in Chapter 1 consisted of a separate D–T-fueled fusion reactor coupled to one or several companion fission reactors by isotope and power exchange linkages. Such a system is called a D–T fusion–fission symbiont. Our objective here is to describe and characterize some of the main features of isotope and power flows for such a fusion–fission nuclear energy system.

6.1. SYSTEM DESCRIPTION

The fusion–fission symbiont system we examine here is taken to be based on the fusion of deuterium nuclei with tritium nuclei in either a magnetic or inertial confinement device. An essential basis we adopt is that a steady-state fusion process characterization can be employed. This does not exclude pulsed operation provided the pulses are sufficiently long and regular to allow an adequate equivalent time-averaged description.

The fusion core is taken to be surrounded by a blanket designed to breed tritium for recycling to the fusion core and fissile fuel for transfer to the companion fission reactor as alluded to in Eq. (3.39). By definition of a symbiont, only negligible fissile fuel burning is to occur in the breeding blanket. Additionally, for reasons of generality, it is assumed that tritium could also be bred in and extracted from the companion fission reactor for transfer to the fusion core. This tritium breeding could

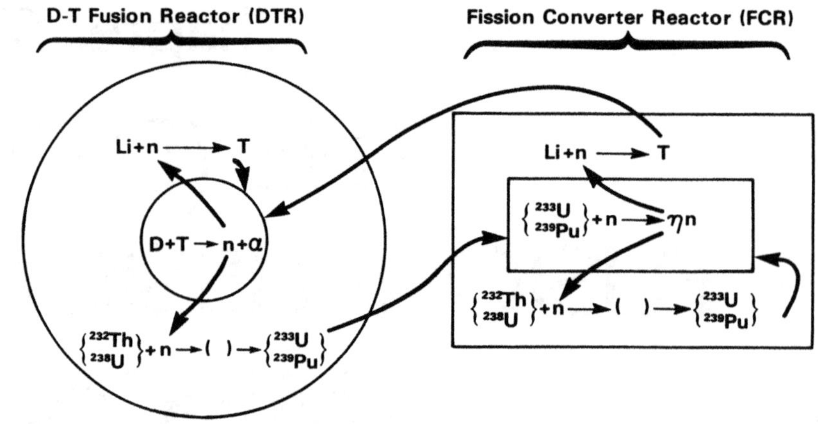

FIGURE 6.1. Schematic illustration of reaction processes and fuel transfer–recycle patterns for the D–T fusion–fission symbiont.

occur incidently in the coolant/moderator or in control rods and related flux-flattening devices containing lithium. Alternatively, one might conceive extensive design provisions in the fission reactor to provide substantial tritium breeding.

In Figure 6.1 we illustrate the various nuclear reaction processes as well as the corresponding fissile and fusile transfer and recycle couplings. As illustrated, this represents a loss-free closed fissile and fusile fuel cycle and assumes steady-state operation, continuous fuel flow, instantaneous fuel transfer, and fuel self-sufficiency. These conditions have a major impact on the analysis of the system and its performance. In this idealized case, the fuel management practice does not affect isotope balance considerations. Other fuel management practices can be incorporated as suggested in Section 5.7.

In our previously defined reactor notation, Section 3.3, such a D–T based fusion–fission symbiont would be represented by

$$\begin{array}{lll} \begin{matrix}[\mathrm{T}] \rightarrow \\ [\mathrm{D}] \rightarrow\end{matrix} & \left\{ \mathrm{D}(\mathrm{T}, {}^4\mathrm{He})n \right\}_{\mathrm{DTC}} & \rightleftarrows P_e \\ \begin{matrix}[\mathrm{Li}] \rightarrow \\ \\ [\mathrm{Fb}] \rightarrow\end{matrix} & \left\{ \begin{matrix}\mathrm{Li}(n, \mathrm{Rp})\mathrm{T} \\ \\ \mathrm{Fb}(n, \beta\gamma)\mathrm{Fi}\end{matrix} \right\}_{\mathrm{DTB}} & \begin{matrix}\rightarrow [\mathrm{T}] \\ \rightleftarrows P_e \\ \rightarrow [\mathrm{Fi}]\end{matrix} \\ \begin{matrix}[\mathrm{Fi}] \rightarrow \\ [\mathrm{Fb}] \rightarrow \\ [\mathrm{Li}] \rightarrow\end{matrix} & \left\{ \begin{matrix}\mathrm{Fb}(n, \beta\gamma)\mathrm{Fi}(n, \mathrm{Rp})\eta_{\mathrm{Fi}} n \\ \\ \mathrm{Li}(n, \mathrm{Rp})\mathrm{T}\end{matrix} \right\}_{\mathrm{FCR}} & \begin{matrix}\rightarrow [\mathrm{T}] \\ \\ \rightarrow P_e\end{matrix} \end{array} \quad (6.1)$$

The fundamental reaction in the fusion reactor core, DTC, is the fusion of a deuterium ion with a tritium ion; this nuclear process and the corresponding reaction rate are represented by

$$\mathrm{T} + \mathrm{D} \rightarrow n + {}^{4}\mathrm{He}, \qquad R_{\mathrm{DT}} \tag{6.2}$$

The corresponding exoergic reaction in the fission converter reactors, FCR, is fission symbolically described by

$$\mathrm{Fi} + n \rightarrow \nu_{\mathrm{Fi}} n + \mathrm{Rp}, \qquad R_{\mathrm{fi,Fi}} \tag{6.3}$$

The pertinent fissile fuel removal due to fission and neutron capture is described in terms of our previously defined notation as

$$\mathrm{Fi} + n \rightarrow \eta_{\mathrm{Fi}} n + \mathrm{Rp}, \qquad R_{a,\mathrm{Fi}} \tag{6.4}$$

Tritium is bred in the fusion blanket, DTB, and/or in the companion fission reactor, FCR, mainly by the reactions

$$\begin{Bmatrix} {}^{6}\mathrm{Li} \\ {}^{7}\mathrm{Li} \end{Bmatrix} + n \rightarrow \begin{Bmatrix} 0 \\ n' \end{Bmatrix} + {}^{4}\mathrm{He} + \mathrm{T}, \qquad R_{c,\mathrm{Li}} \tag{6.5}$$

which can also be represented as Li(n, Rp)T with the reaction products Rp being an alpha particle or an alpha particle plus a neutron as appropriate. Neutron capture in deuterium and ternary fission are examples of other sources of tritium in fission reactors.

The breeding of fissile fuel in the fusion reactor blanket, DTB, and the fission reactor, FCR, occurs mainly via the sequences

$$\begin{Bmatrix} {}^{232}\mathrm{Th} \\ {}^{238}\mathrm{U} \\ {}^{240}\mathrm{Pu} \end{Bmatrix} + n \rightarrow \begin{Bmatrix} {}^{233}\mathrm{Th} \\ {}^{239}\mathrm{U} \\ {}^{241}\mathrm{Pu} \end{Bmatrix} \xrightarrow{\beta-} \begin{Bmatrix} {}^{233}\mathrm{Pa} \\ {}^{239}\mathrm{Np} \end{Bmatrix} \xrightarrow{\beta-} \begin{Bmatrix} {}^{233}\mathrm{U} \\ {}^{239}\mathrm{Pu} \end{Bmatrix}, \qquad R_{c,\mathrm{Fb}} \tag{6.6}$$

or, more compactly

$$\mathrm{Fb} + n \rightarrow \mathrm{Fi} + \cdots, \qquad R_{c,\mathrm{Fb}} \tag{6.7}$$

The relative breeding characteristics of this D–T-based fusion–fission symbiont require more careful attention than those for stand-alone reactors. Whereas the fissile fuel conversion ratio C_{Fi} of a stand-alone fission reactor is a well-known parameter, we must now incorporate the feature that the fissile fuel can also be bred in the fusion reactor blanket. Additionally, since tritium is also a fuel and can be bred either in the fusion or fission components of the system, it will be useful to quantify tritium breeding—that is, fusile fuel breeding—in a similar manner. Therefore, we define fissile and fusile fuel conversion ratios for each component in terms of the breeding reaction rate of fissile and fusile fuel relative to the dominant fuel destruction rate of that component:

(1) Fissile fuel conversion ratio in the fission converter reactor:

$$C_{\mathrm{Fi,FCR}} = \{R_{c,\mathrm{Fb}}/R_{a,\mathrm{Fi}}\}_{\mathrm{FCR}} \qquad (6.8a)$$

(2) Fusile fuel breeding ratio in the fission converter reactor:

$$C_{\mathrm{T,FCR}} = \{R_{c,\mathrm{Li}}/R_{a,\mathrm{Fi}}\}_{\mathrm{FCR}} \qquad (6.8b)$$

(3) Fissile fuel breeding ratio in the D–T fusion reactor blanket:

$$C_{\mathrm{Fi,DTR}} = \{R_{c,\mathrm{Fb}}/R_{\mathrm{DT}}\}_{\mathrm{DTR}} \qquad (6.8c)$$

(4) Fusile fuel breeding ratio in the *D–T* fusion reactor blanket:

$$C_{\mathrm{T,DTR}} = \{R_{c,\mathrm{Li}}/R_{\mathrm{DT}}\}_{\mathrm{DTR}} \qquad (6.8d)$$

Here $C_{\mathrm{Fi,FCR}}$ is the conventional definition of fission reactor physics, whereas $C_{\mathrm{Fi,DTR}}$ refers to fissile fuel breeding normalized to a fusion reaction. A similar interpretation applies to $C_{\mathrm{T,FCR}}$ while $C_{\mathrm{T,DTR}}$ has the conventional definition used for stand-alone fusion reactors. These breeding ratios are consistent extensions and will be found to be of considerable utility in our analysis.

6.2. ISOTOPE SYSTEMATICS

The fissile and fusile fuel rate equations for the symbiont system follow from a specification of a balance condition involving the fuel production and destruction rates. The tritium inventory will be represented by N_T. We write, with the aid of Figure 6.1, the rate equation for tritium

$$\frac{d}{dt}N_T = \{R_{c,\mathrm{Li}}\}_{\mathrm{DTB}} - \{R_{\mathrm{DT}}\}_{\mathrm{DTC}} + \{R_{c,\mathrm{Li}}\}_{\mathrm{FCR}} \tag{6.9}$$

Using the appropriate breeding ratio definitions, we express this continuity equation in terms of the D–T fusion reaction rate in the fusion component and the fissile consumption rate in the fission reactor, which are the two dominant reaction rates:

$$\frac{d}{dt}N_T = \{(C_T - 1)R_{\mathrm{DT}}\}_{\mathrm{DTR}} + \{C_T R_{a,\mathrm{Fi}}\}_{\mathrm{FCR}} \tag{6.10}$$

For algebraic clarity, the domain to which each term applies is indicated.

The fissile rate equation is obtained in a similar manner

$$\frac{d}{dt}N_{\mathrm{Fi}} = \{R_{c,\mathrm{Fb}} - R_{a,\mathrm{Fi}}\}_{\mathrm{FCR}} + \{R_{c,\mathrm{Fb}}\}_{\mathrm{DTB}} \tag{6.11}$$

where fission reactions in the fusion reactor blanket are assumed to be negligible. Using the conversion ratios then yields

$$\frac{d}{dt}N_{\mathrm{Fi}} = \{(C_{\mathrm{Fi}} - 1)R_{a,\mathrm{Fi}}\}_{\mathrm{FCR}} + \{C_{\mathrm{Fi}}R_{\mathrm{DT}}\}_{\mathrm{DTR}} \tag{6.12}$$

We emphasize here that a symbiont is defined with little fission taking place outside the fission reactor. Therefore the reaction rates associated with fissile materials appear only in the terms related to the converter fission reactor. Further, the bred fuel is taken to be removed immediately. However, a more extended analysis would have to account for the consumption of fissile fuel in the fusion blanket and also for the radioactive decay of tritium and possibly ^{241}Pu. We will pursue this expanded perspective in a later chapter.

The above two fuel rate equations can be written in matrix form as

$$\begin{pmatrix} \frac{d}{dt}N_{\mathrm{Fi}} \\ \frac{d}{dt}N_{\mathrm{T}} \end{pmatrix} = \begin{pmatrix} C_{\mathrm{Fi,FCR}} - 1 & C_{\mathrm{Fi,DTR}} \\ C_{\mathrm{T,FCR}} & C_{\mathrm{T,DTR}} - 1 \end{pmatrix} \begin{pmatrix} R_{a,\mathrm{Fi}} \\ R_{\mathrm{DT}} \end{pmatrix} \tag{6.13}$$

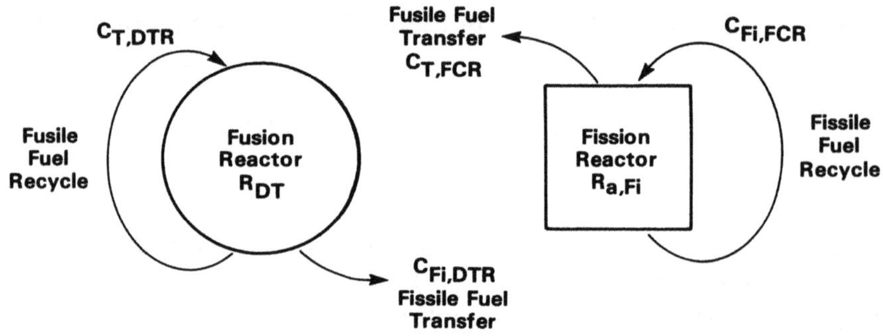

FIGURE 6.2. Symbolic representation of a fusion–fission symbiont illustrating fuel recycle and fuel transfer.

where the diagonal matrix elements describe the fuel flows within the individual reactors and the off-diagonal elements account for the fuel transfers between the components. This equation can be compactly written as

$$\frac{d}{dt}\mathbf{N} = \mathbf{CR} \tag{6.14}$$

where **C** may be interpreted as a fuel coupling matrix linking the symbiont inventory vector **N** to the symbiont reaction rate vector **R**.

Eq. (6.14) describes the comprehensive nuclear fuel coupling possible for a symbiont: both nuclear reactors breed fissile and fusile fuel for transfer and recycle. It may be noted that all of the fuel-coupling combinations need not be considered in order to maintain fuel self-sufficient operation. For example, it might be judged that essential tritium breeding in the fission reactor need not be required, in which case $C_{T,FCR} = 0$. On the other hand, it might be feasible to consider a fission reactor with a low conversion ratio and then require that most of the fissile fuel be supplied by breeding in the fusion blanket.

To enumerate these and other lower-order symbionts, we adopt the notation suggested in Figure 6.2. Here, each of the two coupled reactors is symbolically characterized by the principal reaction rates and the two associated nuclear fuel conversion ratios. With a circle identifying the fusion reactor and a square representing the fission reactor, the various potentially feasible symbionts are distinguished by the presence or absence of the recycle and transfer fuel linkages. A number of isotopically feasible symbionts together with the coupling matrices are illustrated in Figure 6.3. For illustrative and comparative purposes, example No. 1 represents the conventional stand-alone fusion and fission reactors.

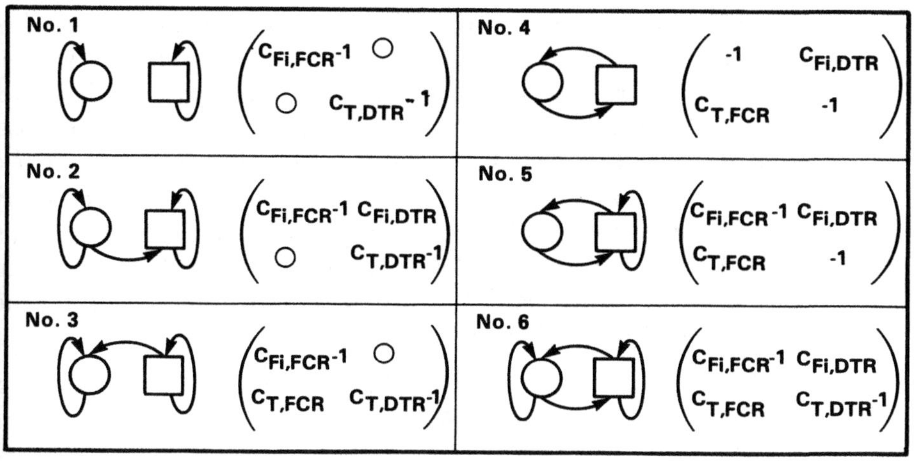

FIGURE 6.3. Listing of several distinct D–T symbionts illustrating the fuel linkages and the associated coupling matrix.

We emphasize that only in the case of a perfectly closed equilibrium fuel cycle may the *in situ* breeding of fissile nuclei in the fission reactor core and the external recycle of fissile fuel contained in fuel elements discharged from the fission reactor be treated identically as suggested by the symbolism employed in Figure 6.2.

6.3. FUEL SELF-SUFFICIENCY

The specification of system parameters of the D–T symbiont can be undertaken at several levels and is dictated by the imposed operational conditions. Our initial intent is to determine the relationships which must hold among the several system parameters under conditions of complete fuel self-sufficiency. That is, following the initial fuel loading of both reactors, the symbiont breeds exactly the amount of fissile and fusile nuclear fuel which it consumes. The implicit assumption is that fuel-cycle equilibrium conditions have been attained.

From a comparison of the graphical isotope flow displayed in Figure 6.1 and the associated isotope rate equations, Eqs. (6.10) and (6.12), we note that complete fuel self-sufficiency is specified by a constant fuel inventory vector **N**, thus giving

$$\frac{d}{dt}N_T = \frac{d}{dt}N_{\mathrm{Fi}} = 0 \tag{6.15}$$

The resultant constraints on the various reaction rates and breedin ratios will depend upon the specific system configuration chosen. We wi not undertake such an analysis for all of the system concepts displaye in Figure 6.3 but select the second system, No. 2, for illustrative purpose

System No. 2 may be labeled a D–T fusion–fission symbiont or D–T assisted fission convertor reactor. The companion fission reactor doe contribute to its own fissile fuel requirement as expressed by $0 < C_{\mathrm{Fi,FCR}}$ 1, but not to the production of fusile fuel for the associated fusion reactc giving $C_{\mathrm{T,FCR}} = 0$. For this case, the steady-state tritium and fissile fue balance equations, Eq. (6.15), result in

$$\{(C_{\mathrm{T}} - 1)R_{\mathrm{DT}}\}_{\mathrm{DTR}} = 0 \qquad (6.1$$

and

$$\{C_{\mathrm{Fi}}R_{\mathrm{DT}}\}_{\mathrm{DTR}} + \{(C_{\mathrm{Fi}} - 1)R_{a,\mathrm{Fi}}\}_{\mathrm{FCR}} = 0 \qquad (6.1$$

yielding the following parametric conditions for this fusion-convert symbiont under condition of fuel self-sufficiency:

$$C_{\mathrm{T,DTR}} = 1 \qquad (6.1$$

and

$$\frac{R_{\mathrm{DT}}}{R_{a,\mathrm{Fi}}} = \frac{1 - C_{\mathrm{Fi,FCR}}}{C_{\mathrm{Fi,DTR}}} \qquad (6.1$$

These equations illustrate concisely the effect of system conversio parameters on the fusile-to-fissile fuel reaction rate ratio. Eq. (6.19) als illustrates the important flexibility in choosing the reaction rate rati $R_{\mathrm{DT}}/R_{a,\mathrm{Fi}}$, and hence the fusion/fission power ratio, through th specification of the design parameters $C_{\mathrm{Fi,DTR}}$ and $C_{\mathrm{Fi,FCR}}$. We will tran late the above reaction rate ratio into a power ratio in the next section.

An examination of the remaining possible D–T symbiont configu ations of Figure 6.3 will demonstrate mathematical similarities amon the various sets of rate equations for this model of fuel self-sufficie operation. Two specific linear forms which relate the reaction rates to th remaining system parameters can be identified as follows:

(1) For those configurations in which there is only one fuel lin between the fusion and the fission component, such as for Nos. 2 and

in Figure 6.3, we obtain linear relationships in the form

$$f_1 R_i = 0 \tag{6.20}$$

and

$$g_1 R_i = R_j \tag{6.21}$$

where f_1 and g_1 are linear functions of the elements of the coupling matrix and where the i-type fuel is the one which is not transferred between the two symbiont components and the j-type fuel is transferred.

(2) For those configurations in which there is a fuel transfer for both types of fuel, we obtain the set

$$f_2 R_i = R_j \tag{6.22}$$

and

$$g_2 R_i = R_j \tag{6.23}$$

Since neither R_i nor R_j is allowed to be zero, the only admissible relationships for the reaction rates and the pertinent self-sufficiency conditions are

$$\frac{R_i}{R_j} = \frac{1}{g_1}, \qquad f_1 = 0 \tag{6.24}$$

for case 1 and

$$\frac{R_i}{R_j} = \frac{1}{g_2}, \qquad f_2 = g_2 \tag{6.25}$$

for case 2. We list the reaction ratios together with the constraints in Table 6.1. Note that these equations are a result of the self-sufficiency imposition, Eq. (6.15), and that the constraints reflect the mathematical condition that two coupled, homogeneous, linear equations have a non-trivial solution only if the determinant of the coupling matrix **C**, Eq. (6.14), vanishes.

TABLE 6.1
Reaction Ratios and Constraints for the D–T Symbiont Configurations Illustrated in Figure 6.3

No.	$R_{DT}/R_{a,FI}$	Constraint
1	(independent)	$C_{Fi,FCR} = C_{T,DTR} = 1$
2	$(1 - C_{Fi,FCR})/C_{Fi,DTR}$	$C_{T,DTR} = 1$
3	$C_{T,FCR}/(1 - C_{T,FCR})$	$C_{Fi,FCR} = 1$
4	$1/C_{Fi,DTR}$	$C_{Fi,DTR}C_{T,FCR} = 1$
5	$(1 - C_{Fi,FCR})/C_{Fi,DTR}$	$C_{Fi,DTR}C_{T,FCR} = 1 - C_{Fi,FCR}$
6	$(1 - C_{Fi,FCR})/C_{Fi,DTR}$	$C_{Fi,DTR}C_{T,FCR} = (1 - C_{Fi,FCR})(1 - C_{T,DTR})$

6.4. CORE AND BLANKET POWER

An itemization of the various power components associated with the fusion core and blanket needs to be undertaken in order to determine the total symbiont station electric power delivered to the grid. We consider here a fusion core energy balance in a steady-state operational mode for which the injected power retained in the plasma is—together with the charged particle fusion power—sufficient to maintain the required ion temperature at a constant level while the heating power not deposited in the plasma is assumed to be unrecoverable. The power balance may then be written as

$$P_I + P_{DT} = P_n + P_r + P_l \tag{6.26}$$

with

$$P_{DT} = P_n + P_\alpha \tag{6.27}$$

The terms are defined as follows:

- P_I: fraction of the externally supplied circulating power $P_{ci,e}$ which is absorbed in the fusion core;
- P_{DT}: nuclear fusion power appearing in the core as a result of the $D + T \rightarrow n + \alpha$ reaction mass decrement;
- P_n: power associated with the kinetic energy of the fusion neutrons passing unimpeded through the fusion core and enter into the blanket;

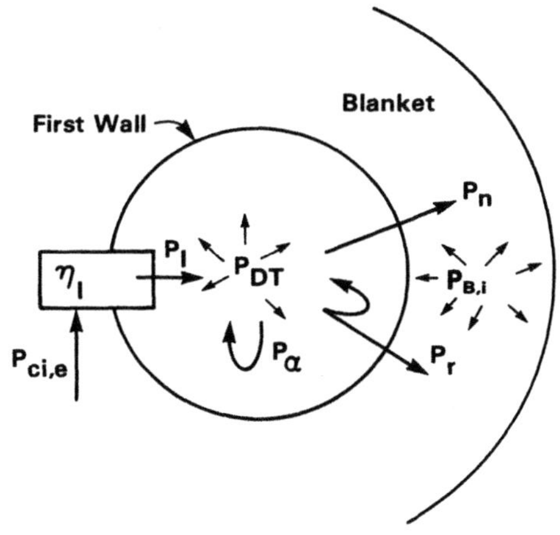

FIGURE 6.4. Power flow components for the fusion reactor.

P_r: fraction of the power associated with bremsstrahlung and cyclotron radiation which is absorbed in or passes through the first wall into the blanket;

P_α: power associated with the kinetic energy of the fusion alpha particle, here taken to be fully retained in the fusion core;

P_l: power losses associated with various processes such as charge exchange, leaking particles, etc.

Equations (6.26) and (6.27) also describe pulsed systems if the terms are defined as suitable averages. As suggested in Figure 6.4, the total extractable thermal power in the blanket is given by

$$P_{B,t} = P_n + P_r + \sum_i P_{B,i} \tag{6.28}$$

The term $P_{B,i}$ refers to the power components associated with the various neutron- and radiation-induced nuclear reactions in the blanket. This may, for example, involve fast fission and photo fission, neutron capture in heavy elements, breeding reactions associated with fissile and fusile fuel, as well as nonfission neutron multiplication.

It is convenient to introduce a blanket power multiplication parameter M_B indicating the relative increase in the blanket power resulting from reactions occurring in the blanket:

$$M_B = \frac{\text{total extractable thermal power in the fusion blanket}}{\text{power from the fusion core entering the first wall}} \tag{6.29}$$

The case of D–T fusion yields specifically

$$M_B = \frac{P_{B,t}}{P_n + P_r}$$

$$= 1 + \frac{\sum_i P_{B,i}}{P_n + P_r} \tag{6.30}$$

and is equal to unity if no power producing reactions are induced in the blanket. This ratio can be viewed as a variable which is a function of the blanket isotopic composition and geometry. We may rewrite Eq. (6.28) therefore in the form

$$P_{B,t} = (P_n + P_r)M_B \tag{6.31}$$

which will find repeated use.

With the main power components specified, we use Figure 6.5 to express the net system electrical power output as

$$P_{o,e} = \eta_{\text{fi},g}P_{\text{fi},t} + \eta_{B,g}P_{B,t} - P_{\text{ci},e} - P_{\text{oc},e} \tag{6.32}$$

where the notation is evident from Figure 6.5 and consistent with the spallation–fission symbiont. By analogy to the spallation–fission symbiont, we first replace the gross thermal-to-electrical conversion efficien-

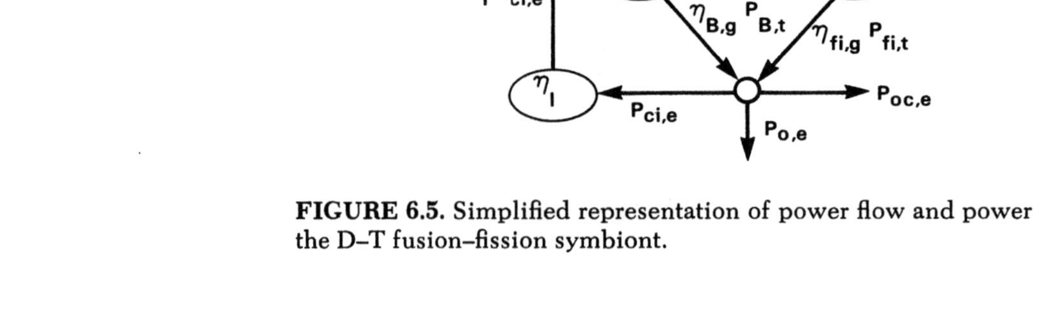

FIGURE 6.5. Simplified representation of power flow and power conversion for the D–T fusion–fission symbiont.

cies $\eta_{\mathrm{fi},g}$ and $\eta_{B,g}$ associated with the heat extracted from the fission reactor and the fusion blanket by the net conversion efficiencies which account for operational and control power requirements. This power is again taken to be a fixed fraction of the thermal power of each of the components, Eq. (4.12):

$$P_{\mathrm{oc},e} = \varepsilon_{\mathrm{fi}} P_{\mathrm{fi},t} + \varepsilon_B P_{B,t} \tag{6.33}$$

The net electrical system power is then

$$\begin{aligned} P_{o,e} &= (\eta_{\mathrm{fi},g} - \varepsilon_{\mathrm{fi}}) P_{\mathrm{fi},t} + (\eta_{B,g} - \varepsilon_B) P_{B,t} - P_{\mathrm{ci},e} \\ &= \eta_{\mathrm{fi}} P_{\mathrm{fi},t} + \eta_B P_{B,t} - P_{\mathrm{ci},e} \end{aligned} \tag{6.34}$$

which may be compared with Eq. (4.14) for the spallation–fission symbiont.

Our intent now is to express the various individual power terms in this equation as a function of both the fission reactor power, $P_{\mathrm{fi},t}$, and the fusile-to-fissile fuel destruction rate ratio, $(R_{\mathrm{DT}}/R_{a,\mathrm{Fi}})$. The latter may then be specified for the condition of fissile and fusile fuel self-sufficiency as listed in Table 6.1. To achieve this we first express the circulating power $P_{\mathrm{ci},e}$ in terms of the fission and fusion reactor power.

The electrical power supplied to the plasma heating devices, $P_{\mathrm{ci},e}$, is related to the power absorbed in the plasma, P_I, by the efficiency η_I with which the supplied electrical power is converted into plasma heat:

$$\eta_I P_{\mathrm{ci},e} = P_I \tag{6.35}$$

Note here that the injection efficiency η_I corresponds to the electrical-to-beam power conversion efficiency η_p in the case of the accelerator symbiont, Eq. (4.20).

To incorporate this input power P_I we introduce a core power multiplication parameter, M_c, according to the discussion in Appendix D and choose it as a figure of merit of plasma performance:

$$\begin{aligned} M_c &= \frac{\text{core power entering the blanket}}{\text{externally supplied power deposited in the core}} \\ &= \frac{P_n + P_r}{P_I} \end{aligned} \tag{6.36}$$

As a companion expression, we introduce a fusion reactor powe multiplication parameter, M_{fu}, by

$$M_{\text{fu}} = \frac{\text{electric power supplied by the fusion reactor}}{\text{electric power required by the fusion reactor}}$$

$$= \frac{\eta_B P_{B,t}}{P_{\text{ci},e}} \tag{6.37}$$

An algebraic manipulation of these M-value expressions and the us of previous relations leads to

$$M_{\text{fu}} = (\eta_I \eta_B M_B) M_c \tag{6.38}$$

The usefulness of Eq. (6.37) is that it allows the explicit expressio of the circulating power $P_{\text{ci},e}$ in terms of the M-value of a fusion reactor

$$P_{\text{ci},e} = \frac{\eta_B P_{B,t}}{M_{\text{fu}}} \tag{6.39}$$

Combining the various expressions yields the station electrical output i terms of the several power components and system parameters of th D–T-based fusion reactor, Eq. (6.34):

$$P_{o,e} = \eta_{\text{fi}} P_{\text{fi},t} + \eta_B (1 - 1/M_{\text{fu}}) P_{B,t} \tag{6.40}$$

which is analogous to Eq. (4.24) for the spallation–fission symbiont an can similarly be written in a more compact form using effective componen conversion ratios:

$$P_{o,e} = \eta_{\text{fi},o} P_{\text{fi},t} + \eta_{\text{fu},o} P_{B,t} \tag{6.41}$$

Here the definitions $\eta_{\text{fi},o}$ and $\eta_{\text{fu},o}$ are evident.

6.5. SYSTEM POWER

In order to relate the fusion component power to the fission-to-fusio rate ratios which are determined by fuel linkage considerations, Sectio

6.3, we rewrite Eq. (6.41) using Eq. (6.31):

$$P_{o,e} = \eta_{\mathrm{fi},o} P_{\mathrm{fi},t} + \eta_{\mathrm{fu},o}(P_n + P_r)M_B \tag{6.42}$$

The power component associated with the kinetic energy of the fusion neutrons, P_n, can be expressed as a function of the fusion power P_{DT} using basic kinematic considerations of the D(T, n)α reaction; this yields both P_n and P_α as

$$P_n = \left\{\frac{m_\alpha}{m_n + m_\alpha}\right\} P_{\mathrm{DT}} \simeq 0.8 P_{\mathrm{DT}} \tag{6.43}$$

$$P_\alpha = \left(\frac{m_n}{m_n + m_\alpha}\right) P_{\mathrm{DT}} \simeq 0.2 P_{\mathrm{DT}} \tag{6.44}$$

where m_α and m_n are the alpha particle and neutron masses, respectively.

Expressing P_r in terms of R_{DT} requires knowledge of various plasma and fusion reactor parameters. For a general parametric analysis, we will find it adequate to use the proportionality parameter γ_r in order to express bremsstrahlung and cyclotron radiation power absorbed in or crossing the first wall:

$$P_r = \gamma_r P_{\mathrm{DT}} \tag{6.45}$$

The evaluation of this radiation-loss-to-fusion-power ratio γ_r is generally complex and involves considerations such as plasma temperature and impurities, magnetic field conditions, wall reflections, absorption in the plasma, particle and power density profiles, as well as other detailed reactor design features. We will therefore consider γ_r to be a design variable.

The station electrical power output can now be compactly written as

$$P_{o,e} = \eta_{\mathrm{fi},o} P_{\mathrm{fi},t} + \eta_{\mathrm{fu},o}(0.8 + \gamma_r) M_B P_{\mathrm{DT}} \tag{6.46}$$

The fission power is related to the fissile fuel consumption rate by Eq. (4.15), with U_{Fi} the energy released per fissile nucleus consumed

$$P_{\mathrm{fi},t} = U_{\mathrm{Fi}} R_{a,\mathrm{Fi}} \tag{6.47}$$

and the fusion power is related to the fusion reaction rate by

$$P_{\mathrm{DT}} = U_{\mathrm{DT}} R_{\mathrm{DT}} \tag{6.48}$$

where U_{DT} is the energy released as a result of a D–T fusion process and has a value of 17.6 MeV (Table 2.2). Therefore, Eq. (6.46) can be written in terms of the fission reactor power and the fusile-to-fissile fuel reaction rate:

$$\begin{aligned} P_{o,e} &= P_{\mathrm{fi},t}\left[\eta_{\mathrm{fi},o} + \eta_{\mathrm{fu},o}(0.8+\gamma_r)M_B\left(\frac{U_{\mathrm{DT}}}{U_{\mathrm{Fi}}}\right)\left(\frac{R_{\mathrm{DT}}}{R_{a,\mathrm{Fi}}}\right)\right] \\ &= P_{\mathrm{fi},t}\left\{\eta_{\mathrm{fi}} + \eta_B[1-(1/M_{\mathrm{fu}})](0.8+\gamma_r)M_B\left(\frac{U_{\mathrm{DT}}}{U_{\mathrm{Fi}}}\right)\left(\frac{R_{\mathrm{DT}}}{R_{a,\mathrm{Fi}}}\right)\right\} \end{aligned} \tag{6.49}$$

With the reaction rates $R_{\mathrm{DT}}/R_{a,\mathrm{Fi}}$ for fuel self-sufficiency operational mode specified in Table 6.1, it is thus possible to calculate the net station power output for a self-sufficient D–T symbiont as a function of the fission reactor power from Eq. (6.49). The various parameters such as M_{fu}, $\eta_{\mathrm{fi},o}$, $\eta_{\mathrm{fu},o}$, etc. are system variables. Thus, the equation is in a convenient form for parametric analyses.

It is apparent from Eq. (6.49) that the symbiont becomes an interesting source of energy even for low plasma power multiplications. However, the degree of its attractiveness is very sensitive to reactor parameters such as the blanket breeding capacity, radiation losses, and the fission reactor conversion ratio, and it strongly depends on energy multiplication.

PROBLEMS

6.1. Formulate the set of reaction rate equations implicit in Eq. (6.1).

6.2. In addition to the five fusion–fission symbionts of Figure 6.3 (Nos. 2–6) identify other combinations based on Eq. (6.13) which do not constitute "useful" systems choices. Can these be characterized by certain properties of the coupling matrix **C**?

6.3. Consider a D–T fusion–fission symbiont which is tritium self-sufficient (No. 2 in Fig. 6.3). Take a current LWR for the fission component ($P_{\mathrm{fi},t} = 3000$ MW, $C_{\mathrm{Fi,FCR}} = 0.5$, $\eta_{\mathrm{fi,0}} = 0.30$) and assume a perfectly closed fuel cycle. For the fusion component assume $\eta_{B,g} = 0.3$, $\gamma_r = 0.2$, $C_{\mathrm{Fi,DTR}} = 1.0$, $M_B = 1.5$ corresponding to a fission suppressed blanket. Calculate the net electric power output of the symbiont, $P_{o,e}$, as a function of the core power multiplication parameter M of the fusion component, M_c.

6.4. Study the trade-off among the various systems parameters for a specified net power output $P_{o,e}$, Eq. (6.49).

5. Plot the results of Problem 6.4, $P_{o,e}$ as a function of M_c, for a variety of parameter values for γ_r and $C_{Fi,DTR}$.
6. Calculate and plot the system efficiency, $\eta_0 = P_{o,e}/(\text{total thermal power})$, as a function of the recirculating power fraction, $\varepsilon_{ci} = P_{ci,e}/P_{o,e}$ for the symbiont defined in Problem 6.4.

Chapter Seven

FISSILE–FUSILE FUEL TRAJECTORIES

The symbiosis of a D–T fusion reactor with a converter fission reactor introduces fuel cycle characteristics which are substantially different from those of a spallation–fission symbiont: both fusion and fission reactors may be directly linked to fissile and to fusile fuel production. Furthermore, the fusile isotope tritium is radioactive, which affects fuel stockpiling for an expanding symbiont economy. The resultant fuel trajectory analysis is therefore expected to be more complex. We derive and examine here the two fuel trajectories, making full use of the stockpile inventory concept.

7.1. SYSTEM CHARACTERIZATION

The schematic characterization of the symbiont of interest here is shown in Figure 7.1, where the fuel-consuming core regions, the fuel-breeding blanket regions, and the corresponding fuel stockpiles are emphasized. The isotope exchange linkages are accordingly indicated. The symbol N_T designates the fusile fuel (i.e., tritium) inventories while N_{Fi} denotes the equivalent fissile fuel inventories (see Appendix C). The graphical depiction in Figure 7.1 notwithstanding, there may or may not be a geometrically distinct breeding blanket associated with the fission reactor. The possibility that the fission component of the symbiont is actually a breeder reactor may be explicitly included, in which case the fusion component contributes to the production of the initial fuel

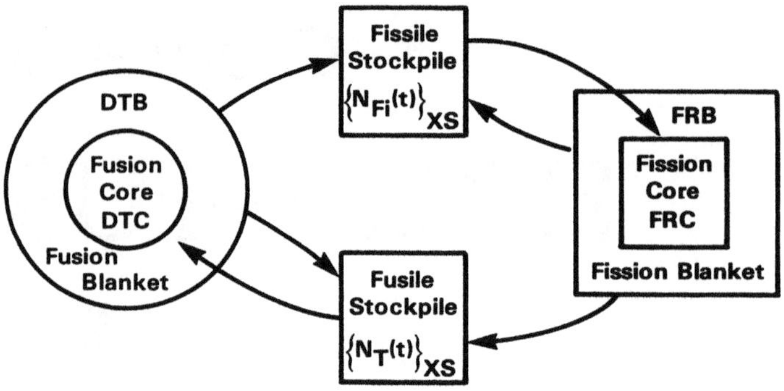

FIGURE 7.1. Schematic depiction of fissile and fusile fuel flow associated with a fully coupled D–T fusion–fission symbiont.

requirement of the fission breeder and enhances the expansion capacity of the energy system. The fission reactor—either converter or breeder—is identified by FR.

Generalizing the analyses of Chapter 5, the stockpile inventory of the ith type of nuclear fuel in its corresponding external stockpile, XS, is defined by a differential relationship written as

$$\left\{\frac{d}{dt}N_i(t)\right\}_{XS} = \left[\frac{d}{dt}N_i(t)\right]_{XS\leftrightarrows iC} + \sum_{jB}\left[\frac{d}{dt}N_i(t)\right]_{XS\leftrightarrows jB} - \lambda_i\{N_i(t)\}_{XS} \tag{7.1}$$

The subscript i represents either fusile or fissile fuel, i = T or Fi. The subscripts iC and jB refer to the core or blanket of either the D–T fusion reactor or the fission reactor, iC = DTC or FRC and jB = DTB or FRB. The { } brackets designate reactions and inventories pertinent to a system component, here DTC, FRC, DTB, FRB, and XS, while the [ ] brackets designate fuel transfer between system components as specified by the subscripts.

The first term on the right-hand side of Eq. (7.1) represents the net fuel flow associated with the two reactor cores while the second term similarly represents contributions to the fuel inventory from the two blanket domains. The last term takes care of the depletion effect due to radioactive decay on the stockpile inventory, with λ_i being the decay constant of the i-type fuel. In both cases of fusile and of fissile fuel, this term can be important since tritium decays with a half-life of $\tau_{1/2} = 12.3$ years and ^{241}Pu with $\tau_{1/2} = 14.4$ years. Our objective now is to solve Eq. (7.1) for $\{N_i(t)\}_{XS}$.

We proceed to define the differential terms in Eq. (7.1) in a manner somewhat similar to that used for the fissile trajectory of the spallation symbiont. Again, we shall develop the formalism for the case of a perfectly closed, loss-free, and instantaneous fuel cycle.

For the core regions, the net fuel flow is taken to be proportional to the net fuel consumption/supply rate

$$\left[\frac{d}{dt}N_i(t)\right]_{XS\rightleftarrows jC} = B_{i,jC}R_i \tag{7.2}$$

where R_i designates the i-type fuel consumption rate in the i-type reactor and $B_{i,jC}$ characterizes the net internal production or consumption of i-type fuel. In contrast to the previously defined conversion ratios $C_{i,j}$ Eq. (6.8), these breeding gains $B_{i,jC}$ describe the net result of production/consumption reactions rather than the gross production. They may therefore be called breeding gains rather than conversion ratios in accordance with customary practice used in the analysis of fission breeder reactors. Note that in Eq. (7.2) we have implied that the fuel residence time in the core is small enough so that radioactive decay need not be included.

For the fusion core, $i = \text{T}$ and $jC = \text{DTC}$. Since tritium breeding occurs only to a negligible extent from parasitic DD fusion reactions in a D–T-based fusion core, we write

$$B_{\text{T,DTC}} \simeq \left\{\frac{0 - R_{\text{DT}}}{R_{\text{DT}}}\right\}_{\text{DTC}} = -1 \tag{7.3}$$

with the tritium consumption rate in the D–T fusion reactor

$$R_i = \{R_{\text{DT}}\}_{\text{DTC}} = R_{\text{DT}} \tag{7.4}$$

There is no need to specify further where these R_{DT} reactions occur since fusion conditions do not exist in the fusion blanket or in the fission reactor. Such considerations will be less trivial in the fission case.

For a conventional fission reactor FR without dedicated fissile breeding domains we have $i = \text{Fi}$, $jC = \text{FR}$

$$B_{\text{Fi,FCR}} = \left\{\frac{R_{c,\text{Fb}} - R_{a,\text{Fi}}}{R_{a,\text{Fi}}}\right\}_{\text{FR}} = C_{\text{Fi,FR}} - 1 < 0 \tag{7.5}$$

and

$$R_i = \{R_{a,\mathrm{Fi}}\}_{\mathrm{FR}} = R_{a,\mathrm{Fi}} \quad (7.6$$

Here, $C_{\mathrm{Fi,FR}}$ is again the conventional conversion ratio, Eq. (6.8a), which represents the number of fissile nuclei bred in the reactor per fissile nucleus consumed.

For a fission reactor FR with a dedicated breeding blanket, the core gain $B_{\mathrm{Fi,FRC}}$ and the blanket breeding gain, $B_{\mathrm{Fi,FRB}}$ must be distinguished. Their sum defines the overall net breeding gain in the fission reactor

$$B_{\mathrm{Fi,FR}} = B_{\mathrm{Fi,FRC}} + B_{\mathrm{Fi,FRB}}$$

$$= \left\{\frac{R_{c,\mathrm{Fb}} - R_{a,\mathrm{Fi}}}{R_{a,\mathrm{Fi}}}\right\}_{\mathrm{FR}} = C_{\mathrm{Fi,FR}} - 1 \quad (7.7$$

Both $B_{\mathrm{Fi,FRC}}$ and $B_{\mathrm{Fi,FRB}}$ are defined as the net number of fissile nuclei bred in the respective reactor zone per fissile nucleus destroyed anywhere in the reactor, thus explicitly including fission and capture events in the blanket:

$$B_{\mathrm{Fi,FRC}} = \frac{\{R_{c,\mathrm{Fb}} - R_{a,\mathrm{Fi}}\}_{\mathrm{FRC}}}{\{R_{a,\mathrm{Fi}}\}_{\mathrm{FR}}}$$

$$= \varepsilon_{\mathrm{Fb},C} C_{\mathrm{Fi,FR}} - \varepsilon_{\mathrm{Fi},C} \quad (7.8$$

and

$$B_{\mathrm{Fi,FRB}} = \frac{\{R_{c,\mathrm{Fb}} - R_{a,\mathrm{Fi}}\}_{\mathrm{FRB}}}{\{R_{a,\mathrm{Fi}}\}_{\mathrm{FR}}}$$

$$= \varepsilon_{\mathrm{Fb},B} C_{\mathrm{Fi,FR}} - \varepsilon_{\mathrm{Fi},B} \quad (7.9$$

Here $C_{\mathrm{Fi,FR}}$ is the conversion/breeding ratio as previously defined, Eq. (6.8a), while $\varepsilon_{\mathrm{Fb},j}$ and $\varepsilon_{\mathrm{Fi},j}$ respectively are the fraction of fissile fuel breeding and destruction processes taking place in the j zone of the fission reactor with $j = C, B$:

$$\varepsilon_{\mathrm{Fb},j} = \frac{\{R_{c,\mathrm{Fb}}\}_j}{\{R_{c,\mathrm{Fb}}\}_{\mathrm{FR}}}, \qquad \sum_j \varepsilon_{\mathrm{Fb},j} = 1 \quad (7.10$$

and

$$\varepsilon_{\mathrm{Fi},j} = \frac{\{R_{a,\mathrm{Fi}}\}_j}{\{R_{a,\mathrm{Fi}}\}_{\mathrm{FR}}}, \qquad \sum_j \varepsilon_{\mathrm{Fi},j} = 1 \tag{7.11}$$

The zonewise definitions of breeding gains become especially simple if one assumes that no fissile fuel destruction takes place outside the core:

$$B_{\mathrm{Fi,FRC}} = \varepsilon_{\mathrm{Fb},C} C_{\mathrm{Fi,FR}} - 1 \tag{7.12}$$

and

$$B_{\mathrm{Fi,FRB}} = \varepsilon_{\mathrm{Fb},B} C_{\mathrm{Fi,FR}} \tag{7.13}$$

The fuel transfer between the fusion core and the external fusile fuel stockpile trajectory is then found to be

$$\left[\frac{d}{dt} N_{\mathrm{T}}(t)\right]_{\mathrm{XS} \rightleftarrows \mathrm{DTC}} = -R_{\mathrm{DT}} \tag{7.14}$$

while the contribution by the fission core to the external fissile fuel stockpile trajectory is

$$\left[\frac{d}{dt} N_{\mathrm{Fi}}(t)\right]_{\mathrm{XS} \rightleftarrows \mathrm{FRC}} = (\varepsilon_{\mathrm{Fi},C} C_{\mathrm{Fb,FR}} - \varepsilon_{\mathrm{Fi},C}) R_{a,\mathrm{Fi}} \tag{7.15}$$

With Eqs. (7.14) and (7.15), the first term of the trajectory equation, Eq. (7.1), has been determined.

7.3. FUEL TRAJECTORY BLANKET COMPONENTS

The blanket breeding contributions will again be characterized by a mean residence time, $\tau_{i,jB}$, of the ith fuel in the jBth breeding region so that the inventory balance equation is a generalization of Eq. (5.25) and becomes

$$\left[\frac{d}{dt} N_i(t)\right]_{\mathrm{XS} \rightleftarrows jB} = \frac{1}{\tau_{i,jB}} \{N_i(t)\}_{jB} \tag{7.16}$$

where $\{N_i(t)\}_{jB}$ is the amount of the ith type of fuel in the jBth breeding region at time t. As suggested earlier, these inventories will vary with time, depending upon the various production and destruction processes:

$$\left\{\frac{d}{dt}N_i(t)\right\}_{jB} = \{(\text{production by breeding})_i - (\text{removal and decay})_i\}_{jB}$$

$$= B_{i,jB}R_j - \frac{1}{\tau_{i,jB}}\{N_i(t)\}_{jB} - \lambda_i\{N_i(t)\}_{jB} \tag{7.17}$$

Here $B_{i,jB}$ is the effective breeding gain of the ith fuel in the jBth blanket, and is defined as the net number of i-type nuclei bred in the jB blanket per j-type nucleus destroyed in the associated j-type reactor.

For tritium, i = T, the effective breeding gain

$$B_{\mathrm{T,DTB}} = \frac{\{R_{c,\mathrm{Li}}\}_{\mathrm{DTB}}}{\{R_{\mathrm{DT}}\}_{\mathrm{DT}}} \simeq \frac{1}{R_{\mathrm{DT}}}\{R_{c,\mathrm{Li}}\}_{\mathrm{DT}}$$

$$= C_{\mathrm{T,DT}} \tag{7.18}$$

gives the net number of T nuclei bred in the fusion blanket per DT fusion reaction and coincides with $C_{\mathrm{T,DT}}$ in Eq. (6.8d). For the fission reactor FR, we obtain an effective tritium gain as

$$B_{\mathrm{T,FRB}} = \frac{\{R_{c,\mathrm{Li}}\}_{\mathrm{FRB}}}{\{R_{a,\mathrm{Fi}}\}_{\mathrm{FR}}} \simeq \frac{1}{R_{a,\mathrm{Fi}}}\{R_{c,\mathrm{Li}}\}_{\mathrm{FR}}$$

$$= C_{\mathrm{T,FR}} \tag{7.19}$$

Furthermore $C_{\mathrm{Fi,DT}}$ is calculated from Eq. (6.8c) for the fusion blanket. For fissile fuel in the fission blanket, $B_{\mathrm{Fi,FRB}}$ was previously determined, Eqs. (7.9) and (7.13) with $C_{\mathrm{Fi,FR}}$ from Eq. (6.8a).

Denoting the initial inventory at $t = 0$ by

$$\{N_i(0)\}_{jB} = N^0_{i,jB} \tag{7.20}$$

and recording the parameters $B_{i,jB}$, $C_{i,jB}$, and $\tau_{i,jB}$ as essentially constant over the time interval of interest, we solve Eq. (7.17) to obtain

$$\{N_i(t)\}_{jB} = \frac{B_{i,jB}\tau_{i,jB}R_j}{1+\lambda_i\tau_{i,jB}}\left\{1 - \left[1 - \frac{N^0_{i,jB}(1+\lambda_i\tau_{i,jB})}{B_{i,jB}\tau_{i,jB}R_j}\right]\exp\left[-t\left(\frac{1}{\tau_{i,jB}} + \lambda_i\right)\right]\right\} \tag{7.21}$$

The asymptotic inventories of the ith fuel in the jBth blanket are given by

$$N_{i,jB}^{\infty} = \lim_{t\to\infty} \{N_i(t)\}_{jB}$$

$$= \frac{B_{i,jB} R_j \tau_{i,jB}}{1 + \lambda_i \tau_{i,jB}} \tag{7.22}$$

Eq. (7.21) may be rewritten in a more compact and convenient form as

$$\{N_i(t)\}_{jB} = N_{i,jB}^{\infty} - (N_{i,jB}^{\infty} - N_{i,jB}^{0}) \exp\left[-t\left(\frac{1}{\tau_{i,jB}} + \lambda_i\right)\right] \tag{7.23}$$

7.4. SYMBIONT FISSILE AND FUSILE FUEL TRAJECTORIES

The differential equation for the inventory of the external stockpile $\{N_i(t)\}_{XS}$, Eq. (7.1), is now given by

$$\left\{\frac{d}{dt} N_i(t)\right\}_{XS} = B_{i,iC} R_i + \sum_{jB} \frac{1}{\tau_{i,jB}} \{N_i(t)\}_{jB} - \lambda_i \{N_i(t)\}_{XS} \tag{7.24}$$

Introduction of Eq. (7.23) for the blanket inventory yields the differential equation

$$\left\{\frac{d}{dt} N_i(t)\right\}_{XS} = B_{i,iC} R_i - \lambda_i \{N_i(t)\}_{XS}$$

$$+ \sum_{jB} \frac{1}{\tau_{i,jB}} \left\{N_{i,jB}^{\infty} - (N_{i,jB}^{\infty} - N_{i,jB}^{0}) \exp\left[-t\left(\frac{1}{\tau_{i,jB}} + \lambda_i\right)\right]\right\} \tag{7.25}$$

At the time of reactor start-up time $t = 0$, the stockpile inventory displays a discontinuity which represents the removal of the initial inventories of the cores, iC, and the blankets, jB, of the system components:

$$\{N_i(0)\}_{XS} = N_{i,XS}^{0+} = N_{i,XS}^{0} - \left(N_{i,iC}^{0} + \sum_{jB} N_{i,jB}^{0}\right), \qquad t = 0^+ \tag{7.26}$$

With $\{N_i(t)\}_{XS}$ specified at $t = 0$ and with the imposition that $B_{i,iC}$ also be a constant, the associated differential equation can readily be solved. The solution for the tritium trajectory is given by

$$\{N_{\mathrm{T}}(t)\}_{\mathrm{XS}} = (N^0_{\mathrm{T,XS}} - N^0_{\mathrm{T,DTC}})\exp(-\lambda_{\mathrm{T}}t) - R_{\mathrm{DT}}\left[\frac{1-\exp(-\lambda_{\mathrm{T}}t)}{\lambda_{\mathrm{T}}}\right] + \frac{N^{\infty}_{\mathrm{T,DTB}}}{\tau_{\mathrm{T,DTB}}}$$

$$\cdot\left[\frac{1-\exp(-\lambda_{\mathrm{T}}t)}{\lambda_{\mathrm{T}}}\right] - N^{\infty}_{\mathrm{T,DTB}}\exp(-\lambda_{\mathrm{T}}t)\left[1-\exp\left(\frac{-t}{\tau_{\mathrm{T,DTB}}}\right)\right]$$

$$+\frac{N^{\infty}_{\mathrm{T,FRB}}}{\tau_{\mathrm{T,FRB}}}\cdot\frac{1-\exp(-\lambda_{\mathrm{T}})}{\lambda_{\mathrm{T}}}$$

$$- N^{\infty}_{\mathrm{T,FRB}}\exp(-\lambda_{\mathrm{T}}t)\left[1-\exp\left(\frac{-t}{\tau_{\mathrm{T,FRB}}}\right)\right] \tag{7.27}$$

Clearly there will be no initial tritium inventory in the blankets and hence $N^0_{\mathrm{T},jB} = 0$. The asymptotic inventories appearing in Eq. (7.27) can be expressed in terms of the previously defined conversion ratios, $C_{\mathrm{T},jB}$, Eq. (6.8). Using Eqs. (7.18), (7.19), and (7.22), we find for the equilibrium tritium inventory in the fusion reactor blanket

$$N^{\infty}_{\mathrm{T,DTB}} = \lim_{t\to\infty}\{N_{\mathrm{T}}(t)\}_{\mathrm{DTB}} = \frac{C_{\mathrm{T,DT}}R_{\mathrm{DT}}}{1/\tau_{\mathrm{T,DTB}} + \lambda_{\mathrm{T}}} \tag{7.28}$$

and, in the fission reactor blanket

$$N^{\infty}_{\mathrm{T,FRB}} = \lim_{t\to\infty}\{N_{\mathrm{T}}(t)\}_{\mathrm{FRB}} = \frac{C_{\mathrm{T,FR}}R_{a,\mathrm{Fi}}}{1/\tau_{\mathrm{T,FRB}} + \lambda_{\mathrm{T}}} \tag{7.29}$$

These equilibrium inventories are associated with the equilibrium transfer rates from the blankets to the stockpile, and the mean fuel residence times, $\tau_{\mathrm{T},jB}$, by

$$T^{\infty}_{\mathrm{T},jB} = [N_{\mathrm{T}}(t)]_{\mathrm{XS}\rightleftarrows jB} = \frac{N^{\infty}_{\mathrm{T},jB}}{\tau_{\mathrm{T},jB}} \tag{7.30}$$

The fissile trajectory is similarly found to be given by

$$\{N_{\mathrm{Fi}}(t)\}_{\mathrm{XS}} = N^0_{\mathrm{Fi,XS}} - (N^0_{\mathrm{Fi,FRC}} + N^0_{\mathrm{Fi,DTB}} + N^0_{\mathrm{Fi,FRB}})$$

$$+ (\varepsilon_{\mathrm{Fb},C}C_{\mathrm{Fi,FR}} - \varepsilon_{\mathrm{Fi},C})R_{a,\mathrm{Fi}}t$$

$$+\frac{N^{\infty}_{\mathrm{Fi,DTB}}}{\tau_{\mathrm{Fi,DTB}}}t-(N^{\infty}_{\mathrm{Fi,DTB}}-N^{0}_{\mathrm{Fi,DTB}})\left[1-\exp\left(\frac{-t}{\tau_{\mathrm{Fi,DTB}}}\right)\right]$$

$$+\frac{N^{\infty}_{\mathrm{Fi,FRB}}}{\tau_{\mathrm{Fi,FRB}}}t-(N^{\infty}_{\mathrm{Fi,FRB}}-N^{0}_{\mathrm{Fi,FRB}})\left[1-\exp\left(\frac{-t}{\tau_{\mathrm{Fi,FRB}}}\right)\right] \quad (7.31)$$

The asymptotic fissile fuel inventory in the fusion blanket, can be derived from Eq. (7.22)

$$N^{\infty}_{\mathrm{Fi,DTB}}=C_{\mathrm{Fi,DT}}R_{\mathrm{DT}}\tau_{\mathrm{Fi,DTB}} \quad (7.32)$$

Similarly, we find for the fission reactor blanket, Eq. (7.9)

$$N^{\infty}_{\mathrm{Fi,FRB}}=[\varepsilon_{\mathrm{Fb},B}C_{\mathrm{Fi,FR}}-\varepsilon_{\mathrm{Fi},B}]R_{a,\mathrm{Fi}}\tau_{\mathrm{Fi,FRB}} \quad (7.33)$$

The pertinent net equilibrium transfer rates are calculated, therefore, according to

$$T^{\infty}_{\mathrm{Fi},jB}=[N_{\mathrm{Fi}}(t)]_{\mathrm{XS}\rightleftarrows jB}=\frac{N^{\infty}_{\mathrm{Fi},jB}}{\tau_{\mathrm{Fi},jB}} \quad (7.34)$$

In Eq. (7.31), we have made provision for initial blanket inventories. Furthermore, here we have specified that the fissile fuel does not undergo radioactive decay; if such a decay would have to be included, particularly in view of the fissile isotope ^{241}Pu, then Eq. (7.31) would have to be replaced by an equation of the type of Eq. (7.27), containing λ_{Fi}. Figure 7.2 provides some general representations of $N_{\mathrm{T}}(t)$, Eq. (7.27), and of $N_{\mathrm{Fi}}(t)$, Eq. (7.31), to illustrate the important and distinct trajectories of these fusile and fissile fuels.

7.5. TRITIUM TRAJECTORY FORMULATION

It is informative to explore some general characteristics of the nonlinear trajectory of fusile fuel, Eq. (7.27), and of fissile fuel, Eq. (7.31). The equations describe the amount of fuel in the two associated stockpiles, Figure 7.1, and, as we indicated, consist of three components: (i) the fuel initially placed into the stockpile prior to reactor start-up, (ii) the fuel withdrawn with time as required to maintain a constant power output,

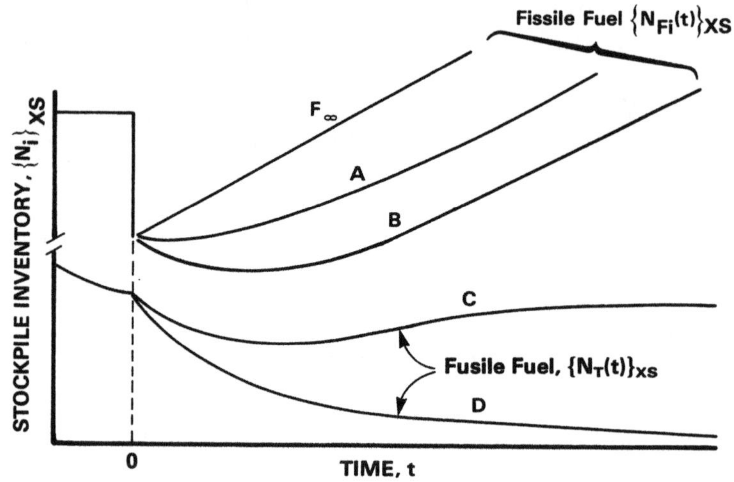

FIGURE 7.2. Fissile fuel trajectories for three cases, each involving the fusion–fission symbiont at steady-state fuel cycle equilibrium.

and (iii) the bred fuel deposited into the stockpile. As described in the derivations, fuel consumption, breeding, hold-up due to residence times in the reactor blankets, and, as appropriate, radioactive decays, are all included, whereas losses and residence times in the fuel cycle, as associated with fuel reprocessing and fabrication, are not included.

While the details of the trajectories will clearly depend upon the specific numerical values of the parameters, the general and dominant features can be conveniently displayed, as shown in Figure 7.2. We comment on the individual curves as displayed.

In order to investigate the preequilibrium features of the trajectories, we first consider the tritium flow and rewrite the fusile fuel trajectory, Eq. (7.27), as

$$\begin{aligned}\{N_T(t)\}_{XS} &= (N^0_{T,XS} - N^0_{T,DTC})\exp(-\lambda_T t)\\ &+ \left[\frac{N^\infty_{T,DTB}}{\tau_{T,DTB}} + \frac{N^\infty_{T,FRB}}{\tau_{T,FRB}} - R_{DT}\right]\frac{1-\exp(-\lambda_T t)}{\lambda_T}\\ &- \left[\frac{N^\infty_{T,DTB}}{\tau_{T,DTB}}\,\frac{1-\exp(-t/\tau_{T,DTB})}{1/\tau_{T,DTB}}\right.\\ &+ \left.\frac{N^\infty_{T,FRB}}{\tau_{T,FRB}}\,\frac{1-\exp(-t/\tau_{T,FRB})}{1/\tau_{T,FRB}}\right]\exp(-\lambda_T t) \qquad (7.35)\end{aligned}$$

By extension of the previously introduced notation this can be written and discussed in a more compact form:

$$\{N_T(t)\}_{XS} = N_{T,XS}^{0+} f(\lambda_T t) + T_T^{\infty} g(\lambda_T t) - \sum_{jB} T_{T,jB}^{\infty} g\left(\frac{t}{\tau_{T,jB}}\right) f(\lambda_T t) \quad (7.36)$$

Here $N_{T,XS}^{0+}$ is the stockpile inventory at time $t = 0^+$, that is, just after the removal of the initial inventory for the fusion core, Eq. (7.26). Furthermore, T_T^{∞} is the net equilibrium tritium transfer rate from the fusion and fission components to the stockpile and the $T_{T,jB}^{\infty}$ are the equilibrium transfer rates from the jB-blanket to the stockpile, Eq. (7.30). Most importantly, the time functions f and g describe the competing inventory buildup and removal processes. Specifically

$$f(\kappa t) = \exp(-\kappa t) = \begin{cases} 1, & t = 0 \\ 1 - \kappa t, & \kappa t \ll 1 \\ 0, & \kappa \tau \gg 1 \end{cases} \quad (7.37)$$

describes a removal procedure, here radioactive decay, which affects inventories. The associated differential equation is

$$\frac{d}{dt} f(\kappa t) = -\kappa f(\kappa t) \quad (7.38)$$

The latter function, in a slight modification of the previous definition, Eq. (5.51)

$$g(\kappa t) = \frac{1 - \exp(-\kappa t)}{\kappa} = \begin{cases} 0, & t = 0 \\ t, & \kappa t \ll 1 \\ 1/\kappa, & \kappa t \gg 1 \end{cases} \quad (7.39)$$

describes the net fuel accumulation as a result of constant production rate competing with an inventory dependent removal rate—here radioactive decay and fuel extraction from blankets. This is evident from the defining equation

$$\frac{d}{dt}g(\kappa t) = [1 - \kappa g(\kappa t)] = f(\kappa t) \qquad (7.4C$$

In view of this itemization, we recognize in Eq. (7.27), or in a mor compact form, Eq. (7.36), the following processes: (i) the first tern describes the removal of the stockpile inventory via radioactive decay (ii) the second term represents the competition between constant ne accumulation in the stockpile from equilibrium supply/consumptio processes, and (iii) the third term accounts for the competition betwee constant accumulation in the blankets from breeding reactions and th removal from inventory-proportional supply to the stockpile superim posed with radioactive decay.

The asymptotic behavior of the functions f and g, Eqs. (7.37) an (7.39), can serve to characterize the stockpile inventory time-depender features, that is, the preequilibrium, equilibrium, and asymptoti domains. The first interval, $0 < t < \tau_{\mathrm{T},jB}$, is of interest for the time perio following reactor start-up and displays a quadratic time dependence o the preequilibrium tritium trajectory with a negative slope and a positiv curvature:

$$\{N_{\mathrm{T}}(t)\}_{\mathrm{XS}} = N^{0+}_{\mathrm{T,XS}} - [R_{\mathrm{DT}} + \lambda_{\mathrm{T}} N^{0+}_{\mathrm{T,XS}}]t + \left(\sum_{jB} T^{\infty}_{\mathrm{T},jB}\right)\lambda_{\mathrm{T}} t^2$$

$$0 \le t < \tau_{\mathrm{T},jB} < \lambda_{\mathrm{T}} \qquad (7.4$$

Once the blanket inventories have been built up, which occurs fo $t \gg \tau_{\mathrm{T},jB}$, the symbiont operates in its equilibrium regime, for which w find from Eq. (7.36)

$$\{N_{\mathrm{T}}(t)\}_{\mathrm{XS}} = [N^{0}_{\mathrm{T,XS}} - (N^{0}_{\mathrm{T,DTC}} + N^{\infty}_{\mathrm{T,DTB}} + N^{\infty}_{\mathrm{T,FRB}})]f(\lambda_{\mathrm{T}} t) + T^{\infty}_{\mathrm{T}} g(\lambda_{\mathrm{T}} t)$$

$$t \gg \tau_{\mathrm{T},jB} \qquad (7.42$$

This suggests revising Eq. (7.36) with $N^{0}_{\mathrm{T},jB} \neq 0$ to a more explicit forn

$$\{N_{\mathrm{T}}(t)\}_{\mathrm{XS}} = \left[N^{0}_{\mathrm{T,XS}} - \left(N^{0}_{\mathrm{T,DTC}} + \sum_{jB} N^{\infty}_{\mathrm{T},jB}\right)\right] f(\lambda_{\mathrm{T}} t)$$

$$+ T^{\infty}_{\mathrm{T}} g(\lambda_{\mathrm{T}} t) + \sum_{jB} (N^{\infty}_{\mathrm{T},jB} - N^{0}_{\mathrm{T},jB}) f(t/\tau_{\mathrm{T},jB}) f(\lambda_{\mathrm{T}} t) \qquad (7.43$$

Here the last term accounts for the preequilibrium feature and vanishes for $t \gg \tau_{T,jB}$ thus merging Eq. (2.43) into Eq. (7.42).

By extension to our previous analysis, Chapter 5, we recognize that the long-term fuel requirement of a symbiont is adequately described by assuming that the equilibrium blanket fuel inventories are externally supplied at the time of reactor startup

$$N^{0}_{i,jB} = N^{\infty}_{i,jB} \tag{7.44}$$

With the previous initial condition of Eq. (7.26) replaced by

$$\{N_i(0)\}_{XS} = N^{0+}_{i,XS} = N^{0}_{i,XS} - \left(N^{0}_{i,iC} + \sum_{jB} N^{\infty}_{i,jB}\right) \tag{7.45}$$

the last term in Eq. (7.43) vanishes, yielding Eq. (7.42) as the complete solution for the trajectory for all $t > 0$.

We conclude that the long-term fuel supply/requirement history of a symbiont is independent of whether the equilibrium blanket inventories are externally supplied or bred *in situ*. As we have seen this is true also in the case of radioactive fuel. One may therefore use a quasidynamic model in which the actual initial inventory condition is replaced by Eq. (7.45) and in which the equilibrium description Eq. (7.42) is used for all $t \geq 0^{+}$.

Eventually, for $t \gg \tau_{1/2}$, the tritium trajectory becomes time independent with its asymptotic value given by

$$N^{\infty}_{T,XS} = \lim_{t\to\infty} \{N_T(t)\}_{XS} = \frac{1}{\lambda_T}\left[\frac{N^{\infty}_{T,DTB}}{\tau_{T,DTB}} + \frac{N^{\infty}_{T,FRB}}{\tau_{T,FRB}} - R_{DT}\right] \tag{7.46}$$

The asymptotic inventory is therefore linked to the equilibrium net fuel transfer rate between the symbiont components and the stockpile according to

$$T^{\infty}_{T} = \lambda_T N^{\infty}_{T,XS} \tag{7.47}$$

and depends on the blanket fuel management parameters, $\tau_{T,jB}$, and on the burn rate, R_{DT}, but is independent of initial fuel provisions from the stockpile. Since this asymptotic inventory must be positive, we impose the condition

$$R_{DT} \leq \sum_{jB} \frac{N^{\infty}_{T,jB}}{\tau_{T,jB}} \tag{7.48}$$

and insertion of Eqs. (7.28) and (7.29) yields therefore

$$\left(\frac{C_{T,DT}}{1+\tau_{T,DTB}\lambda_T} - 1\right)R_{DT} + \left(\frac{C_{T,FR}}{1+\tau_{T,FRB}\lambda_T}\right)R_{a,Fi} \geqslant 0 \tag{7.49}$$

which represents the asymptotic tritium self-sufficiency requirement; the fusile fuel production must make up for consumption and decay of the fuel. Equation (7.49) also specifies the lower bound for the fission-to-fusion power ratio associated with a given set of conversion and fuel management parameters $C_{T,jB}$ and $\tau_{T,jB}$, respectively:

$$\frac{R_{a,Fi}}{R_{DT}} \geqslant \left(\frac{1+\lambda_T\tau_{T,FRB}}{1+\lambda_T\tau_{T,DTB}}\right)\left(\frac{1+\lambda_T\tau_{T,DTB} - C_{T,DT}}{C_{T,FR}}\right) \tag{7.50}$$

With this result we may finally rewrite the trajectory, Eq. (7.36), in a particularly tractable form as

$$\{N_T(t)\}_{XS} = N^{\infty}_{T,XS} + \left[(N^{0+}_{T,XS} - N^{\infty}_{T,XS}) - \sum_{jB} T^{\infty}_{T,jB}\, g(t/\tau_{T,jB})\right] f(\lambda_T t) \tag{7.51}$$

which is more general than its derivation suggests if $g(t)$ is taken to be determined by all preequilibrium features of the system; $g(t)$ clearly reduces to the simple form given by Eq. (7.39) if the preequilibrium behavior of the symbiont is solely due to the blanket inventory built up. For instantaneous fuel removal from the blankets $\tau_{T,jB} = 0$, we find from Eqs. (7.28), (7.29), and (7.46)

$$N^{\infty}_{T,XS} = (1/\lambda_T)[(C_{T,DT} - 1)R_{DT} + C_{T,FR}R_{a,Fi}] \tag{7.52}$$

and from Eq. (7.51)

$$\{N_T(t)\}_{XS} = [(C_{T,DT} - 1)R_{DT} + C_{T,FR}R_{a,Fi}]\frac{1-\exp(-\lambda_T t)}{\lambda_T} + N^{0+}_{T,XS}\exp(-\lambda_T t) \tag{7.53}$$

We will use these latter formulations in the following where we explore the characteristics of the fusile fuel trajectory further.

7.6. TRITIUM TRAJECTORY PARAMETRIZATION

With Eqs. (7.51)–(7.53) we have thus available a general tritium trajectory representation which lends itself particularly well to further investigation. An especially interesting question pertains to the minimum amount of tritium initially required to start and operate a fusion reactor. Evidently, we must have

$$\{N_{\mathrm{T}}(t)\}_{\mathrm{XS}} \geq 0, \qquad \text{for all } t \geq 0 \tag{7.54}$$

The extremum of the tritium trajectory is obtained by differentiation. In our trajectory representation Eq. (7.51) and for $g(t)$ specified by Eqs. (7.39) and (7.40), the fuel flow can be shown to be characterized by

$$\left\{\frac{d}{dt}N_{\mathrm{T}}\right\}_{\mathrm{XS}} = \left\{-\lambda_{\mathrm{T}}(N_{\mathrm{T,XS}}^{0+} - N_{\mathrm{T,XS}}^{\infty}) - \sum_{jB} T_{\mathrm{T},jB}^{\infty}\left[1 - \frac{1+\lambda_{\mathrm{T}}\tau_{\mathrm{T},jB}}{\tau_{\mathrm{T},jB}} g(t/\tau_{\mathrm{T},jB})\right]\right\} f(\lambda_{\mathrm{T}}t) \tag{7.55}$$

or, more specifically

$$\left\{\frac{d}{dt}N_{\mathrm{T}}\right\}_{\mathrm{XS}} = \{-(\lambda_{\mathrm{T}}N_{\mathrm{T,XS}}^{0+} + R_{\mathrm{DT}}) + R_{\mathrm{DT}}C_{\mathrm{T,DT}}[1 - \exp(-t/\tau_{\mathrm{T,DTB}})] + R_{a,\mathrm{Fi}}C_{\mathrm{T,FR}}[1 - \exp(-t/\tau_{\mathrm{T,FRB}})]\} \exp(-\lambda_{\mathrm{T}}t) \tag{7.56}$$

This shows that following reactor start-up, the stockpile inventory is depleted at the rate

$$\left\{\frac{d}{dt}N_{\mathrm{T}}(t)\right\}_{\mathrm{XS}} = -(\lambda_{\mathrm{T}}N_{\mathrm{T,XS}}^{0+} + R_{\mathrm{DT}}), \qquad t \simeq 0^{+} \tag{7.57}$$

which is in agreement with Eq. (7.41). It then reaches a minimum at th point in time $t_{\mathrm{T},m}$ defined by

$$\left\{\frac{d}{dt}N_{\mathrm{T}}(t)\right\}=0, \qquad t=t_{\mathrm{T},m} \tag{7.5}$$

and finally becomes constant, Eq. (7.46):

$$\lim_{t\to\infty}\left\{\frac{d}{dt}N_{\mathrm{T}}(t)\right\}_{\mathrm{XS}}=0 \tag{7.5}$$

From Eq. (7.58) we calculate $t_{\mathrm{T},m}$ by insertion of Eq. (7.56)

$$R_{\mathrm{DT}}C_{\mathrm{T,DT}}\exp(-t_{\mathrm{T},m}/\tau_{\mathrm{T,DTB}})+R_{a,\mathrm{Fi}}C_{\mathrm{T,FR}}\exp(-t_{\mathrm{T},m}/\tau_{\mathrm{T,FRB}})$$
$$=-(\lambda_{\mathrm{T}}N^{0+}_{\mathrm{T,XS}})+R_{\mathrm{T,DT}}(C_{\mathrm{DT}}-1)+R_{a,\mathrm{Fi}}C_{\mathrm{T,FR}} \tag{7.6}$$

For instantaneous fuel removal from the breeding blankets, $\tau_{\mathrm{T},jB}=0$, th left-hand side of Eq. (7.60) is zero and the condition Eq. (7.58) has solution only for the trivial case of a constant stockpile inventory. Th can be seen by setting the right-hand side of Eq. (7.60) equal to zero an introducing the resultant expression for $N^{0+}_{\mathrm{T,XS}}$ into Eq. (7.56).

For noninstantaneous blanket fuel discharge, the left-hand side Eq. (7.60) assumes only positive values or, in the limit, can be zero if th minimum is reached asymptotically, $t_{\mathrm{T},m}=\infty$. Therefore setting the rig side of Eq. (7.60) equal to zero yields the threshold inventory value

$$(N^{0+}_{\mathrm{T,XS}})_{\mathrm{th}}=\frac{R_{\mathrm{DT}}}{\lambda_{\mathrm{T}}}\left(C_{\mathrm{T,DT}}-1+\frac{R_{a,\mathrm{Fi}}}{R_{\mathrm{DT}}}C_{\mathrm{T,FR}}\right) \tag{7.6}$$

Note that there is no minimum if the actual initial inventory excee this threshold value, Eq. (7.61), even if tritium removal from the blanke is not instantaneous:

$$t_{\mathrm{T},m}=\infty \quad \text{and} \quad \{N_{\mathrm{T}}(t_{\mathrm{T},m})\}_{\mathrm{XS}}=N^{\infty}_{\mathrm{T,XS}} \tag{7.6}$$

n case the actual initial inventory is lower than the threshold value, Eq. 7.61), the minimum stockpile inventory is found from Eq. (7.55) and Eqs. (7.46) and (7.51) as

$$\{N_T(t_{T,m})\}_{XS} = N^{\infty}_{T,XS} - \sum_{jB} \frac{T^{\infty}_{T,jB}}{\lambda_T} f(t_{T,m}/\tau_{T,jB}) f(t_{T,m}\lambda_T) \qquad (7.63)$$

or, again, more specifically

$$\{N_T(t_{T,m})\}_{XS} = N^{\infty}_{T,XS} - \frac{R_{DT}}{\lambda_T}\left\{\frac{C_{T,DT}}{1+\lambda_T\tau_{T,DTB}} \exp\left[-t_{T,m}\left(\frac{1}{\tau_T}+\lambda_T\right)\right]\right.$$

$$\left. + \frac{R_{a,Fi}}{R_{DT}} \cdot \frac{C_{T,FR}}{1+\lambda_T\tau_{T,FRB}} \exp\left[-t_{T,m}\left(\frac{1}{\tau_{T,FRB}}+\lambda_T\right)\right]\right\} \qquad (7.64)$$

An interesting point worth exploring has now appeared. We know that $t_{T,m}$ increases with increasing $N^0_{T,XS}$ until $t_{T,m} = \infty$ for $N^0_{T,XS} = (N^0_{T,XS})_{crit}$, Eq. (7.61). Concurrently the minimum stockpile inventory itself also increases until it reaches $N^{\infty}_{T,XS}$, Eq. (7.62). Since there are only monotonous functions involved, the opposite also holds and the minimum inventory decreases with decreasing initial inventory. This is of particular interest since, in general, one should aim to have as small an initial external tritium stockpile inventory as possible. However, there is clearly a lower limit to the initial stockpile inventory $N^0_{T,XS}$ which is determined by the imposition that the symbiont tritium requirement must be met for all t, Eq. (7.54). Thus the smallest $N^0_{T,XS}$ is that for which the minimum stockpile inventory is zero. Therefore

$$\{N_T(t)\}_{XS} = \left\{\frac{d}{dt}N_T(t)\right\}_{XS} = 0, \qquad t = t^0_{T,m} = \text{Min: } \{t_{T,m}\} \qquad (7.65)$$

defines the smallest possible $t_{T,m}$, here designated by $t^0_{T,m}$, and, concurrently, the lowest possible amount of tritium required by the symbiont from external sources, $(N^0_{T,XS})_m$.

We find $t^0_{T,m}$ by setting the minimum stockpile inventory, Eqs. (7.63) and (7.64), equal to zero. Introducing this $t^0_{T,m}$ into the minimum condition, Eq. (7.60), then defines the lowest possible value for $N^{0+}_{\mathrm{T,XS}}$:

$$(N^{0+}_{\mathrm{T,XS}})_m = \{N^0_{\mathrm{T}}(t^0_{\mathrm{T},m})\}_{\mathrm{XS}}$$

$$= \frac{R_{\mathrm{DT}}}{\lambda_{\mathrm{T}}}\left[-1 + \sum_{jB} \frac{T^{\infty}_{\mathrm{T},jB}}{R_{\mathrm{DT}}}(1 + \lambda_{\mathrm{T}}\tau_{\mathrm{T},jB})\frac{g(t^0_{\mathrm{T},m}/\tau_{\mathrm{T},jB})}{\tau_{\mathrm{T},jB}}\right]$$

$$= \frac{R_{\mathrm{DT}}}{\lambda_{\mathrm{T}}}\left\{-1 + C_{\mathrm{T,DT}}[1 - \exp(-t^0_{\mathrm{T},m}/\tau_{\mathrm{T,DTB}})]\right.$$

$$\left. + \frac{R_{a,\mathrm{Fi}}}{R_{\mathrm{DT}}} C_{\mathrm{T,FRB}}[1 - \exp(-t^0_{\mathrm{T},m}/\tau_{\mathrm{T,DTB}})]\right\} \tag{7.66}$$

As illustrated in Figure 7.3, we have thus found that as $N^{0+}_{\mathrm{T,XS}}$ decreases, the trajectory first has no minimum and no inflection point for finite points in time, curve *A*. As $N^{0+}_{\mathrm{T,XS}}$ attains a value below the threshold value $(N^{0+}_{\mathrm{T,XS}})_{\mathrm{th}}$ given by Eq. (7.61), the trajectory displays a minimum for $t_{\mathrm{T},m} < \infty$ while the asymptotic value $N^{\infty}_{\mathrm{T,XS}}$ remains unchanged, curve *B* in Figure 7.3. As $N^{0+}_{\mathrm{T,XS}}$ continues to decrease, it finally reaches the lower bound given by Eq. (7.66) for which the stockpile is completely depleted when its minimum is reached. Thus $(N^{0+}_{\mathrm{T,XS}})_m$, Eq. (7.66), is the minimum initial tritium inventory required—in addition to $N^0_{\mathrm{T,DTC}}$—to

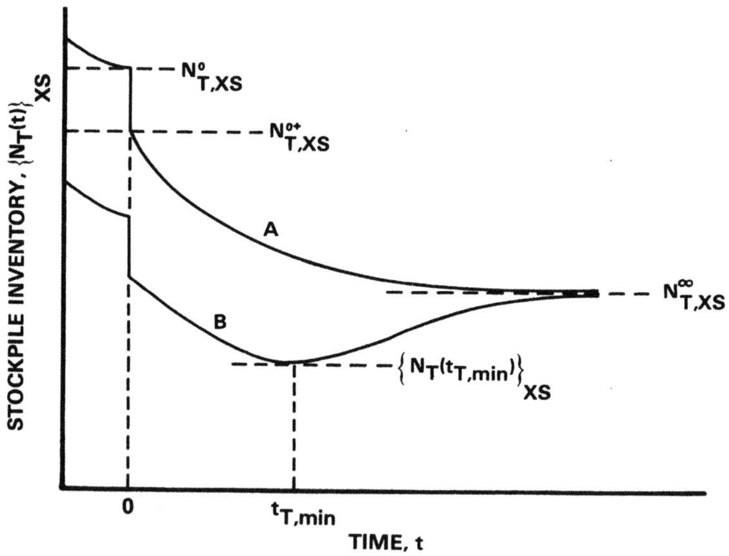

FIGURE 7.3. Detailed schematic illustration of symbiont tritium trajectory for the case of high (a) and low (b) initial tritium inventories.

operate the reactor until $t = t_{T,m}$; from then on more tritium becomes available than is required for the reactor operation.

As an illustrative example we calculate this minimum initial stockpile inventory requirement for the case that tritium is bred only in the fusion blanket. We find from Eq. (7.64)

$$\exp(-t^0_{T,m}/\tau_{T,DTB}) = \left(\frac{\lambda_T N^{\infty}_{T,XS}}{R_{DT}} \cdot \frac{1+\lambda_T \tau_{T,DTB}}{C_{T,DT}}\right)^{1/(1+\lambda_T \tau_{T,DTB})}$$

$$= \left(1 - \frac{1+\lambda_T \tau_{T,DTB}}{C_{T,DT}}\right)^{1/(1+\lambda_T \tau_{T,DTB})} \quad (7.67)$$

Inserting this into Eq. (7.66) yields

$$(N^0_{T,XS})_m = N^0_{T,XS}(t^0_{T,m})$$

$$= N^0_{T,DTC} + \frac{R_{DT}}{\lambda_T}\left\{C_{T,DT} - 1 - C_T\left(1 - \frac{1+\lambda_T \tau_{T,DTB}}{C_{T,DT}}\right)^{1/(1+\lambda_T \tau_{T,DTB})}\right. \quad (7.68)$$

as the minimum initial tritium stockpile requirement.

7.7. FISSILE FUEL TRAJECTORY CHARACTERIZATION

We now consider some additional characteristics of the fissile fuel trajectory, Eqs. (7.31) to (7.34). It is convenient to do this again with the previously defined functions $f(t)$ and $g(t)$, Eqs. (7.37) to (7.40):

$$\{N_{Fi}(t)\}_{XS} = N^{0+}_{Fi,XS} f(\lambda_{Fi} t) + T^{\infty}_{Fi} g(\lambda_{Fi} t)$$

$$- \left[\sum_{jB} [T^{\infty}_{Fi,jB} - T^0_{Fi,jB}] g(t/\tau_{Fi,jB}) f(\lambda_{Fi} t)\right] \quad (7.69)$$

If all fuel isotopes of interest are stable, then $\lambda_{Fi} \to 0$, $f(\lambda_{Fi} t) \to 1$, $g(\lambda_{Fi} t) \to t$, and we obtain

$$\{N_{Fi}(t)\}_{XS} = N^{0+}_{Fi,XS} + T^{\infty}_{Fi} t - \sum_{jB} (T^{\infty}_{Fi,jB} - T^0_{Fi,jB}) g(t/\tau_{Fi,jB}) \quad (7.70)$$

Apparently, the equilibrium ($t \gg \tau_{\mathrm{Fi},jB}$) and asymptotic ($t \rightarrow \infty$) tim domains converge for the case of stable fuel. Thus Eq. (7.70) correspond to both Eq. (7.42) and Eq. (7.46) in the case of tritium. The first term i Eq. (7.70) again describes the stockpile inventory after supplying th initial core and blanket inventories,

$$N_{\mathrm{Fi,XS}}^{0+} = N_{\mathrm{Fi,XS}}^{0} - N_{\mathrm{Fi,FRC}}^{0} - \sum_{jB} N_{\mathrm{Fi},jB}^{0} \qquad (7.71)$$

The second term in Eq. (7.70) is the net asymptotic fissile fuel transfe between symbiont and stockpile

$$\begin{aligned} T_{\mathrm{Fi}}^{\infty} &= \sum_{jB} T_{\mathrm{Fi},jB}^{\infty} + [\varepsilon_{\mathrm{Fb,C}} C_{\mathrm{Fi,FR}} - \varepsilon_{\mathrm{Fi,C}}] R_{a,\mathrm{Fi}} \\ &= C_{\mathrm{Fi,DT}} R_{\mathrm{DT}} + [C_{\mathrm{Fi,FR}} - 1] R_{a,\mathrm{Fi}} \end{aligned} \qquad (7.72)$$

Finally, the third term accounts for the inventory build-up in the blanket The term $T_{\mathrm{Fi},jB}^{\infty}$ characterizes the equilibrium fissile fuel transfer rat from the blankets jB to the stockpile, Eq. (7.34), and $T_{\mathrm{Fi},jB}^{0}$ describe th corresponding initial fissile fuel flows

$$T_{\mathrm{Fi},jB}^{0} = \frac{N_{\mathrm{Fi},jB}^{0}}{\tau_{\mathrm{Fi},jB}} \qquad (7.73)$$

Considering the fissile fuel trajectories, Eq. (7.70), and curves (A and (B) of Figure 7.2, we note that for a sufficiently large t, the exponentia terms vanish, leaving a linear asymptote:

$$\begin{aligned} \{N_{\mathrm{Fi}}(t)\}_{\mathrm{XS}} &= [N_{\mathrm{Fi,XS}}^{0} - (N_{\mathrm{Fi,FRC}}^{0} + N_{\mathrm{Fi,DTB}}^{\infty} + N_{\mathrm{Fi,FRB}}^{\infty})] \\ &\quad + [(C_{\mathrm{Fi,FR}} - 1) R_{a,\mathrm{Fi}} + C_{\mathrm{Fi,DT}} R_{\mathrm{DT}}] t, \qquad t \gg \tau_{\mathrm{Fi},jB} \end{aligned} \qquad (7.74)$$

The slope of the asymptotic trajectory is fully determined by the net fissil fuel production rate. The sign of the coefficient of t determines whethe the system asymptotically requires externally supplied fissile fuel. If thi coefficient is positive, implying

$$\frac{R_{\mathrm{DT}}}{R_{a,\mathrm{Fi}}} > \frac{1 - C_{\mathrm{Fi,FR}}}{C_{\mathrm{Fi,DT}}} \qquad (7.75)$$

then the system is a net breeder of fissile fuel. This imposition is to be compared with Eq. (7.50) for the fusile fuel which emerged from the imposition that the asymptotic breeder inventory be a positive quantity. Both Eq. (7.50) and Eq. (7.75) can therefore be considered as important conditions under which the symbiont provides its own fissile and fusile fuel and defines, for a given set of parameters, the range of possible reaction ratios. For the case that the fuel is radioactive, it is apparently advantageous to keep the fissile stockpile inventory always to a minimum. Hence, Eq. (7.65) is to be included in a fuel analysis assuring the external fuel supply required to reach the self-sufficiency operational mode, Eqs. (7.46) to (7.50), with a minimum amount of fuel.

Another important feature can be recognized in Eq. (7.74). It is evident that supplying an initial inventory to the breeding zones may accelerate the eventual achievement of the asymptotic behavior but will not affect the extent to which the asymptote is displaced with respect to the idealized zero-blanket-inventory case designated in Figure 7.2 by F_∞. We have encountered the same effect for the tritium trajectory, Eq. (7.42), for the interesting intermediate time interval. As suggested in Eqs. (7.42)–(7.45) we can adequately describe the long-term fissile fuel history without a detailed reference to preequilibrium features if we replace the actual initial condition, Eq. (7.71), by

$$N^{0+}_{\mathrm{Fi,XS}} = N^{0}_{\mathrm{Fi,XS}} - N^{0}_{\mathrm{Fi,FRC}} - \sum_{jB} N^{\infty}_{\mathrm{Fi},jB} \tag{7.76}$$

and then assume the validity of the equilibrium formulation, Eq. (7.74), from reactor startup. This follows along the line of Eqs. (7.42) to (7.68).

The limiting case F_∞, Figure 7.2, corresponds to an instantaneous extraction of bred fuel from all zones:

$$\lim_{\tau_{\mathrm{Fi},jB}\to 0} \{N_{\mathrm{Fi}}(t)\}_{\mathrm{XS}} = N^{0}_{\mathrm{Fi,XS}} - N^{0}_{\mathrm{Fi,FRC}} + [C_{\mathrm{Fi,DT}}R_{\mathrm{DT}} + (C_{\mathrm{Fi,FR}} - 1)R_{a,\mathrm{Fi}}]t, \qquad t \geq 0 \tag{7.77}$$

and represents a common, though obviously inadequate, description of the fuel flow adopted in practice for breeding systems. We emphasize that this idealized case points to the importance of the explicit or implicit consideration of preequilibrium system features for a proper system analysis, Eq. (7.70) or Eq. (7.72), respectively.

Trajectory *B* of Figure 7.2 also illustrates the effect of increasing mean residence times $\tau_{\mathrm{Fi,DTB}}$ and $\tau_{\mathrm{Fi,FRB}}$. Increasing these residence times increases external supply of additional fissile fuel until a net production of fissile fuel is attained. The time at which the minimum inventory occurs is again found by setting $\{dN_{Fi}/dt\}_{XS} = 0$, which in general yields a transcendental equation for $t_{Fi,min}$:

$$\left\{\frac{d}{dt}N_{Fi}\right\} = T_{Fi}^{\infty} - \sum_{jB}[T_{Fi,jB}^{\infty} - T_{Fi,jB}^{0}]f(t/\tau_{Fi,jB}) = 0, \qquad t = t_{Fi,min} \tag{7.78}$$

A simple algebraic equation for $t_{Fi,m}$ is possible if both blankets possess the same residence time, $\tau_{Fi,DTB} = \tau_{Fi,FRB} = \tau_{Fi,B}$ yielding

$$t_{Fi,m} = \tau_{Fi,B} \ln \frac{\sum_{jB}(B_{Fi,jB}R_{jB} - N_{Fi,jB}^{0}/\tau_{Fi,B})}{\sum_{jB}(B_{Fi,jB}R_{jB}) + B_{Fi,FRC}R_{a,Fi}} \tag{7.79}$$

Similar expressions result if one of the mean residence times is essentially zero or if breeding occurs only in one blanket such as in the case of a converter reactor without a proper blanket, $B_{Fi,FRB} = 0$; then is

$$\begin{aligned} t_{Fi,m} &= \tau_{Fi,DTB} \ln\left(\frac{B_{Fi,DTB}R_{DT} - N_{Fi,DTB}^{0}/\tau_{Fi,DTB}}{B_{Fi,DTB}R_{DT} + B_{Fi,FRC}R_{a,Fi}}\right) \\ &= \tau_{Fi,DTB} \ln\left(\frac{C_{Fi,DT}R_{DT} - N_{Fi,DTB}^{0}/\tau_{Fi,DTB}}{C_{Fi,DT}R_{DT} + (\varepsilon_{Fb,C}C_{Fi,FRC} - \varepsilon_{Fi,C})R_{a,Fi}}\right) \end{aligned} \tag{7.80}$$

In the limit of $\tau_{Fi,DT} = \tau_{Fi,FRB} \to 0$, no transient depression occurs and the minimum fissile inventory occurs at the time of reactor start-up, curve F_∞ in Figure 7.2 and Eq. (7.77). In all other cases the stockpile inventory at $t = t_{Fi,min}$ can be calculated by inserting Eq. (7.79) into Eq. (7.70) thus defining the inventory of the stockpile at the point in time of its maximal depletion,

$$\begin{aligned} \{N_{Fi}(t_{Fi,m})\}_{XS} = {} & N_{Fi,XS}^{0+} + T_{Fi}^{\infty}t_{Fi,m} \\ & - \sum_{jB}[T_{Fi,jB}^{\infty} - T_{Fi,jB}^{0}]g(t_{Fi,m}/\tau_{Fi,jB}) \end{aligned} \tag{7.81}$$

Clearly the essential requirement is that

$$\{N_{Fi}(t_{Fi,m})\}_{XS} \geq 0 \tag{7.82}$$

which defines the lower bound of fissile fuel that has to be made externally available in order to start and operate a fission reactor in a fusion–fission symbiont until it eventually becomes either fuel self-sufficient or even a net fissile fuel producer: Specifically, this gives this lower bound by

$$(N^0_{\mathrm{Fi,XS}})_m = N^0_{\mathrm{Fi,FRC}} - T^\infty_{\mathrm{Fi}} t_{\mathrm{Fi},m} + \sum_{jB} [T^\infty_{\mathrm{Fi},jB} - T^0_{\mathrm{Fi},jB}] g(t_{\mathrm{Fi},m}/\tau_{\mathrm{Fi},jB}) \qquad (7.83)$$

with all parameters as defined previously.

PROBLEMS

7.1. Generalize the trajectory analysis to $k = 1, 2, \ldots, K$ distinct breeding regions.
7.2. Identify the conditions that satisfy the requirement for a D–T symbiont to be asymptotically self-sufficient with respect to both fissile and fusile fuel.
7.3. Specify the reactor parameters for $\{N_{\mathrm{Fi}}(T)\}_{\mathrm{XS}}$ such that $(d/dt)\{N_{\mathrm{Fi}}(t)\}_{\mathrm{XS}} = 0$ at $t = 0$ and interpret its implications.
7.4. Calculate the point parameters such as $t_{\mathrm{T,min}}$; $N^\infty_{\mathrm{T,XS}}$; $T^\infty_{\mathrm{T,XS}}$; $(N^0_{\mathrm{T,XS}})_m$, etc. with the tritium trajectory of a fusion reactor. For the initial inventory assume one value above, one value below the threshold value, Eq. (7.61); for $\tau_{\mathrm{T,DTB}}$ take 1 month and 1 year, for other parameters assume typical values.
7.5. Plot the tritium trajectory $\{N_{\mathrm{T}}(t)\}_{\mathrm{XS}}$ with $\tau_{\mathrm{T,DTB}}$ as parameter, e.g., $\tau_{\mathrm{T,DTB}} = 1$ day, 1 month, 1 year and discuss the results.
7.6. Discuss the origin of the qualitative difference between tritium and fissile trajectories, Figure 7.2.

Chapter Eight

THE (D–T) FUSION–FISSION HYBRID

A symbiont system has been defined in previous chapters as consisting of two separate though coupled nuclear facilities. In contrast, a hybrid reactor is one nuclear power station but based on two distinct and dominant nuclear processes. In particular, the D–T hybrid consists of a D–T-burning fusion core surrounded by a blanket which displays many characteristics of a subcritical fission reactor core and which also breeds tritium. The fusion-source neutrons from the core combine with the fission-source neutrons to establish a steady-state neutron population in this blanket. Though the hybrid clearly differs fundamentally from a symbiont in both concept and technological requirements, similar analytical representations for both are possible, allowing therefore a degree of continuity with previous chapters.

8.1. SYSTEM DESCRIPTION

The distinguishing feature of a D–T hybrid is that the core is surrounded by a fission reactor type blanket of some convenient geometry: spherical, cylindrical, or toroidal. Though $T(D,\alpha)n$ will be the only reaction of relevance in the core, a number of distinct neutron-induced reactions must be accounted for in the blanket. Figure 8.1 provides a schematic depiction of these various processes. The blanket reactions and reaction rates of interest here can be enumerated as follows: The D–T

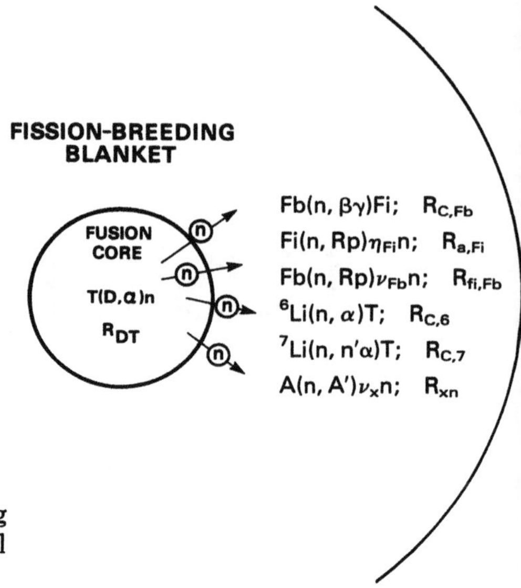

FIGURE 8.1. Illustration and listing of blanket reactions and fissile fuel flow of a (D–T) hybrid.

fusion rate occurring in the reactor core is represented by

$$D + T \rightarrow n + {}^{4}He, \qquad R_{DT} \tag{8.1}$$

The emitted 14.1-MeV neutrons enter the blanket and contribute to the neutron population. The blanket may be viewed as a subcritical fission assembly driven by an external high-energy neutron source. Thermal and fast fission of fissile nuclei will be represented by $R_{fi,Fi}$ while the total fissile fuel consumption rate, consisting of fission and capture, will again be described by $R_{a,Fi}$:

$$Fi + n \rightarrow \nu_{Fi} n + \sum_i Rp_i, \qquad R_{fi,Fi} \tag{8.2}$$

and

$$Fi + n \rightarrow \eta_{Fi} n + \sum_i Rp_i, \qquad R_{a,Fi} \tag{8.3}$$

Fissile fuel breeding is represented by the neutron capture rate $R_{c,Fb}$ by fertile material in the blanket:

$$Fb + n \rightarrow Fi + (\cdots), \qquad R_{c,Fb} \tag{8.4}$$

This bred fissile fuel may be used as fuel makeup in the blanket or as fuel for an external fission reactor economy.

In addition, fusile breeding processes are necessary to supply the fusion core with tritium. Breeding via neutron absorption in lithium is expected to be the dominant tritium source

$$^{6}\text{Li} + n \rightarrow {}^{4}\text{He} + \text{T}, \qquad R_{c,6} \tag{8.5}$$

and

$$^{7}\text{Li} + n_f \rightarrow n + {}^{4}\text{He} + \text{T}, \qquad R_{c,7} \tag{8.6}$$

where n_f again designates a fast neutron.

In order to obtain a more comprehensive accounting of the neutron balance, neutron multiplying processes such as inelastic (n, xn) reactions in the blanket and fast fission reactions of nonfissile heavy nuclei are taken into account. These neutron-multiplying reactions can be represented by

$$A + n_f \rightarrow xn + A', \qquad R_{xn,A}, \qquad x = 2, 3, \ldots \tag{8.7}$$

and

$$\text{Fb} + n \rightarrow \nu_{\text{Fb}} n + \sum_i \text{Rp}_i, \qquad R_{\text{fi,Fb}} \tag{8.8}$$

where A' represents reaction products of mass $A + 1 - x$. Note that in Eqs. (8.2) and (8.8), the fission neutron yield is defined separately for a fissile and a fertile nucleus.

By decomposing the neutron-multiplying reactions, an expression for a space and energy average neutron multiplication yield in the blanket is obtained:

$$\nu_x = \frac{1}{R_{xn}} \int_B \int_E [2\Sigma_{n,2n}(E, \mathbf{x}) + 3\Sigma_{n,3n}(E, \mathbf{x}) + \cdots]\phi(E, \mathbf{x})\, d\mathbf{x}\, dE \tag{8.9}$$

Here $\Sigma_{n,xn}(E, \mathbf{x})$ is the space and neutron energy dependent macroscopic cross section for inelastic (n, xn) reactions induced by fast neutrons in all isotopes present in the blanket:

$$\Sigma_{n,xn}(E, \mathbf{x}) = \sum_A \sigma_{(n,xn),A}(E) N_A(\mathbf{x}) \tag{8.10}$$

This effective neutron multiplication yield, Eq. (8.9), is associated with the overall neutron multiplication rate:

$$R_{xn} = \int_B \int_E [\Sigma_{n,2n}(E, \mathbf{x}) + \Sigma_{n,3n}(E, \mathbf{x}) + \cdots] \phi(E, \mathbf{x}) \, dE \, d\mathbf{x} \tag{8.11}$$

Finally, parasitic neutron capture reactions and neutron losses by escape are additional terms needed to complete a detailed balance description.

In terms of the reactor notation of Section 3.4, the D–T hybrid can be represented by the following scheme, which also emphasizes the predominant neutron flows:

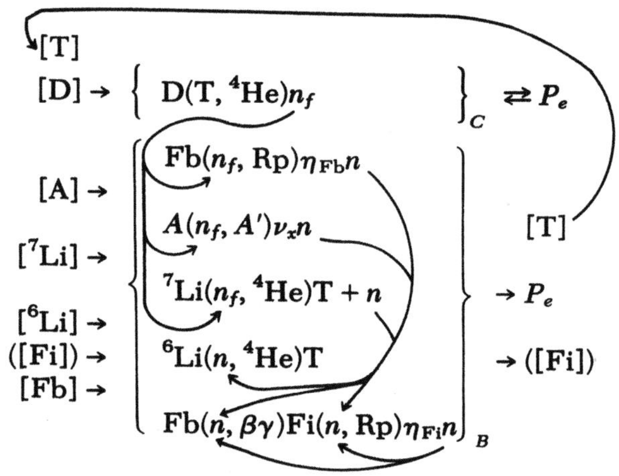

(8.12)

Fusion neutron and fission energy, as well as energy released in other processes, is deposited in the blanket for eventual conversion into electrical form. Of the several previously employed blanket parameters, there are two which are particularly pertinent in hybrid reactor analysis. One is the total fuel breeding capacity of the hybrid and the other is the blanket power multiplication. The former identifies the total number of fissile and fusile fuel nuclei which are bred in the blanket for each fusion reaction while the latter refers to the extent to which the 14-MeV fusion neutron energy is multiplied in the blanket. These terms are analytically defined as needed in subsequent sections.

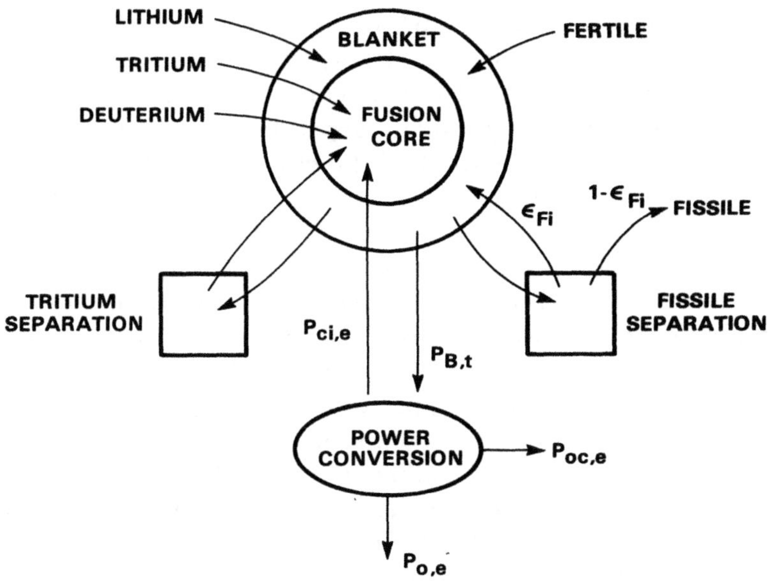

FIGURE 8.2. Power flow and isotope flow options of the D–T hybrid.

8.2. ISOTOPE FLOW OPTIONS

The various power and isotope feed flows which may be required are illustrated in Figure 8.2. These external feeds are characterized by the following distinct operational modes which will have an important influence on the fission/breeding blanket design:

(i) fusile–fertile fuel (^{6}Li, ^{7}Li) is externally supplied to the blanket or, alternatively, fusile fuel tritium ($^{3}H = T$) is introduced from external sources to the core;
(ii) fissile–fertile fuel (^{232}Th, ^{238}U) is externally supplied to the blanket;
(iii) fusile fuel deuterium ($^{2}H = D$) is externally supplied to the core.

For our purpose here, we make an important distinction between the supply of fissile fuel and fusile fuel. We require that the fissile fuel, that is ^{233}U, ^{239}Pu, or ^{241}Pu, be supplied to the blanket only by recycle following isotopic separation, while the fusile fuel tritium may be provided to the fusion core by a combination of external supply and internal recycle from its production by (n, α) or $(n, n'\alpha)$ reactions in lithium located in the blanket. The external tritium supply could be associated with dedicated or specially designed fission reactors using lithium as control poisons or with extraction of accumulated tritium from reactor cooling water where it is produced by incidental (n, γ) reactions. In terms of our previous terminology then, the hybrid here is conceived to be either (i) fully or partially self-sufficient or (ii) totally dependent upon external fusile fuel sources. With respect to fissile fuel, however, the hybrid is taken to be either totally self-sufficient or even a net breeder of fissile fuel.

Fissile fuel flow may be characterized by the fraction ε_{Fi} of bred fissile fuel burned *in situ*. Then the fraction $(1-\varepsilon_{Fi})$ of bred fissile fuel is the net amount of fissile fuel transferred to an external stockpile for eventual use by other fission reactors. Introducing the parameter ε_{Fi} allows for the convenient description of the case of fissile fuel self-sufficiency as well as the case of breeding for an external fission reactor economy. For steady-state isotope flow conditions $\varepsilon_{Fi}=1$ represents fuel self-sufficiency while $0<\varepsilon_{Fi}<1$ defines the hybrid as a fuel breeder. The choice of $\varepsilon_{Fi} \simeq 0$ will not be considered here since, in a sense, it represents a symbiont studied in Chapters 6 and 7.

As previously indicated, we will concern ourselves first with several distinct fusion–fission systems concepts which will be identified by the following acronyms: ^{6}Li hybrid, ^{7}Li hybrid, and T hybrid. These three systems differ by the method of tritium supply which defines three distinct operational modes of the hybrid.

(i) ^{6}*Li hybrid*: ^{6}Li is the only externally supplied fertile fuel (Figure 8.3a).

(ii) ^{7}*Li hybrid*: ^{7}Li is the only externally supplied fertile fuel (Figure 8.3b).

(iii) *T hybrid*: Tritium is supplied externally thus eliminating the need for tritium breeding in the fusion reactor blanket (Figure 8.3c).

Note that in this specific analysis, the fission fuel flows—both fertile and fissile—are taken to be identical according to the equivalent fuel concept, Appendix C, for each of the three systems. Only the management of the fusion fuels ^{6}Li, ^{7}Li, and T differs. Additionally, the blankets for each of these modes of operation would evidently be of different design.

From the graphical similarity of Figures 8.3a and 8.3b it is tempting to assign similar operational characteristics to the ^{6}Li hybrid and ^{7}Li hybrid. As will become clear, the role of the $(n, n'\alpha)$ cross section in ^{7}Li is significant. Indeed, if there is any similarity between any of these three systems, it will be found that this applies to the ^{7}Li hybrid and T hybrid. We will examine each of these systems separately using an approach which differs slightly from the analysis used for the D–T symbiont.

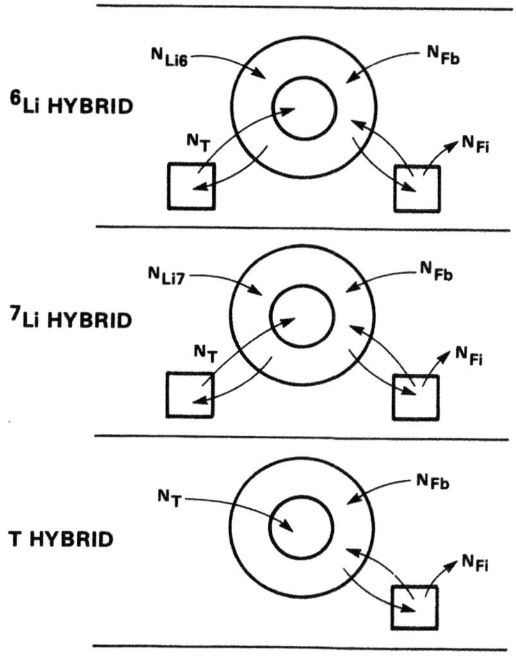

FIGURE 8.3. Illustration showing the differences between three D–T hybrid concepts.

8.3. THE SELF-SUFFICIENT ^{6}Li HYBRID

The isotope flows for the ^{6}Li hybrid are displayed in Figure 8.3. The choice $\varepsilon_{Fi} = 1$ describes a truly self-sufficient system. The case of the hybrid as a fissile fuel breeder will be discussed in the next section.

In contrast to our previous description of the inventory dynamics for the fissile, N_{Fi}, and fusile fuels, N_T, we will find it convenient to establish also a balance equation for the neutron population, N_n. This additional equation will be useful in the formulation of particularly compact relationships and will circumvent the need to introduce additional reactor physics parameters. Evidently, the neutron lifetime will be dominated by absorption processes so that the radioactive lifetime of unbound neutrons need not be incorporated in the analysis.

Though we are dealing with three distinct hybrid reactor types, we can nevertheless identify a subset of reaction rates common to all three reactor types. This includes the D–T fusion rate, Eq. (8.1), the fissile fuel fission and consumption rate, Eqs. (8.2) and (8.3), respectively, the fissile breeding rate, Eq. (8.4), and the neutron multiplication and fertile fission rate, Eqs. (8.7) and (8.8), respectively. The only reaction rate which is system specific is the rate at which tritium is bred from ^{6}Li:

$$R_{c,6} = \int_R \int_E \Sigma_{c,6}(E, \mathbf{x})\phi(E, \mathbf{x})\, dE\, d\mathbf{x} \tag{8.13}$$

Using the above reaction rates, we write the rate equation for the ^{6}Li hybrid with the aid of Figure 8.3a for the fissile fuel, fusile fuel, and the neutrons as follows:

$$\frac{d}{dt}N_T = R_{c,6} - R_{DT} \tag{8.14}$$

$$\frac{d}{dt}N_{Fi} = \varepsilon_{Fi}R_{c,Fb} - R_{a,Fi} \tag{8.15}$$

and

$$\frac{d}{dt}N_n = R_{DT} + (\eta_{Fi} - 1 - \delta_n)R_{a,Fi} - R_{c,6} - R_{c,Fb} + (\nu_x - 1)R_{xn} + (\nu_{Fb} - 1)R_{fi,Fb} \tag{8.16}$$

Note that fissile fuel self-sufficiency is associated here with $\varepsilon_{Fi} = 1$.

In the last equation, the right-hand side describes the various neutron producing and consuming reactions. The first two terms give the fusion and fission neutron source strength, δ_n is the average neutron loss due to parasitic processes expressed as a fraction of the fissile fuel consumption rate, η_{Fi} is the neutron energy and space averaged number of neutrons emitted per fissile nucleus consumed, and ν_x and ν_{Fb} are the neutron yield of nonfission neutron multiplication and fertile fission reactions, respectively.

Note that in Eq. (8.15) the possibility that a fast-neutron-induced inelastic scattering reaction of the type $A(n_f, xn)A'$ destroys fissile fuel if $A = \text{Fi}$, or produces fissile nuclei if $A' = \text{Fi}$, is not included because of their relative infrequent occurrence. Note also that Eqs. (8.14) and (8.15) are based on the assumption that all fuel transfers and reloadings are instantaneous and loss free thus eliminating the need for a specific accounting in a fuel management program which might include tritium recovery as well as feed and fissile fuel recycle.

Imposing steady-state operational conditions, the rate equations for the ^{6}Li hybrid, Eqs. (8.14)–(8.16), are subjected to

$$\frac{d}{dt}N_T = \frac{d}{dt}N_{Fi} = \frac{d}{dt}N_n = 0 \tag{8.17}$$

giving specifically the following set of identities involving reaction rates and reactor parameters:

$$R_{c,6} = R_{\mathrm{DT}} \tag{8.18}$$

$$R_{c,\mathrm{Fb}} = R_{a,\mathrm{Fi}}, \quad \varepsilon_{\mathrm{Fi}} = 1 \tag{8.19}$$

$$R_{\mathrm{DT}} + (\eta_{\mathrm{Fi}} - 1 - \delta_n)R_{a,\mathrm{Fi}} = R_{c,6} + R_{c,\mathrm{Fb}} - (\nu_x - 1)R_{xn} - (\nu_{\mathrm{Fb}} - 1)R_{\mathrm{fi,Fi}} \tag{8.20}$$

This set of three equations possesses four neutron spectrum and system design dependent parameters, η_{Fi}, δ_n, ν_x, and ν_{Fb}, and six reaction rates. Algebraic manipulation of these equations leads to several useful results.

As an example we consider a relationship giving the neutron multiplication requirement for fissile and fusile fuel self-sufficiency in terms of the number of nonfission neutron multiplication reactions per fissile nucleus consumed. For this the above equations yield

$$\frac{R_{xn}}{R_{a,\mathrm{Fi}}} = \frac{2 - \eta_{\mathrm{Fi}} + \delta_n - (\nu_{\mathrm{Fb}} - 1)f}{\nu_x - 1} \tag{8.21}$$

where we have again used the fertile fission fraction f defined as the number of fast fission reactions of fertile nuclei per fissile nucleus consumed. This relation, Eq. (8.21), indicates that for a fast neutron spectrum involving large values for η_{Fi}, ν_{Fb}, and f the numerator could become negative for small neutron losses δ_n; however, avoiding that, since the reaction rates have to represent positive quantities, this ^{6}Li hybrid would of necessity be a breeder, in which case $\varepsilon_{\mathrm{Fi}} < 1$ would apply. In all other cases the (n, xn) reactions would have to balance the neutron economy as specified by this equation.

The important observation to note is that the 14-MeV fusion neutrons are more suited to induce neutron multiplication reactions including fast fission than the fission neutrons, whereas the slowing down fission neutrons are more effective in inducing breeding reactions in ^{6}Li as well as fissile fission reactions. Hence the important aim is to improve the neutron balance by coupling the reactions, Eq. (8.21), and by making full use of the multiplication capacity of fast neutrons as suggested by the following:

$$D + T \rightarrow n + \alpha, \qquad Fi + n \rightarrow \eta_{fi} n + Rp \tag{8.22}$$

neutron multiplication and breeding

This may be achieved in practice by proper design of the blanket geometry and space-dependent isotopic composition. The implications of these considerations thus point to the importance of the neutron spectrum and neutron multiplication in the blanket, which is more evident next when we consider the ^{6}Li hybrid as a fissile fuel breeder.

8.4. THE ^{6}Li-HYBRID BREEDER

Some characteristics of the ^{6}Li hybrid as a breeder of fissile fuel may be determined by imposing steady-state conditions, Eq. (8.17), on the pertinent rate equations, Eqs. (8.14) to (8.16) subject to $0 < \varepsilon_{Fi} < 1$. While this defines a breeder with respect to fissile fuel, it nevertheless is still taken to be a self-sufficient system with respect to its fusile fuel isotope tritium. The characterizing equations for the ^{6}Li hybrid are therefore

$$R_{c,6} = R_{DT} \tag{8.23}$$

$$\varepsilon_{Fi} R_{c,Fb} = R_{a,Fi} \tag{8.24}$$

and

$$R_{DT} + (\eta_{Fi} - 1 - \delta_n) R_{a,Fi} = R_{c,6} + R_{c,Fb} - (\nu_x - 1) R_{xn} - (\nu_{Fb} - 1) R_{fi,Fb} \tag{8.25}$$

Eliminating the breeding reaction rates $R_{c,6}$ and $R_{c,Fb}$ yields an equation involving fission and neutron multiplication together with the various system parameters and the fissile recycle fraction ε_{Fi}

$$(\eta_{Fi} - 1) + (\nu_x - 1)\frac{R_{xn}}{R_{a,Fi}} + (\nu_{Fb} - 1)\frac{R_{fi,Fb}}{R_{a,Fi}} = \delta_n + \frac{1}{\varepsilon_{Fi}} \tag{8.26}$$

Defining

$$\eta_o = \eta_{\mathrm{Fi}} + (\nu_x - 1)\frac{R_{xn}}{R_{a,\mathrm{Fi}}} + (\nu_{\mathrm{Fb}} - 1)\frac{R_{\mathrm{fi,Fb}}}{R_{a,\mathrm{Fi}}} \tag{8.27}$$

as the overall number of neutrons made available through fissile and fertile fuel fission and neutron multiplication per fissile nucleus consumed, we obtain a compact expression for the fissile fuel recycle fraction which maintains steady-state operation:

$$\varepsilon_{\mathrm{Fi}} = \frac{1}{\eta_o - 1 - \delta_n} \tag{8.28}$$

Additionally, since we have assumed that no fissile fuel is supplied from external sources, $\varepsilon_{\mathrm{Fi}} \lesssim 1$, we further require

$$\eta_o - \delta_n \geq 2 \tag{8.29}$$

This relation is similar to one encountered in fission reactor physics except for the fact that η_o here is much larger. This is due to the higher neutron yield per fission, the lower capture-to-fission ratio, and the larger contribution of fertile fission and neutron multiplications in the harder neutron spectrum of a fusion–fission blanket.

The usefulness of Eq. (8.28) arises from its close connection to the fissile fuel breeding ratio of the ^{6}Li hybrid. Using the conventional definition of a fissile breeding ratio and denoting this ratio by $C_{\mathrm{Fi},6}$ for the ^{6}Li hybrid reactor gives

$$C_{\mathrm{Fi},6} = \frac{\text{net fissile fuel production rate}}{\text{fissile fuel destruction rate}}$$

$$= \frac{R_{c,\mathrm{Fb}}}{\varepsilon_{\mathrm{Fi}} R_{c,\mathrm{Fb}}} = \frac{1}{\varepsilon_{\mathrm{Fi}}} \tag{8.30}$$

where $R_{c,\mathrm{Fb}}$ is the production rate of fissile fuel and $\varepsilon_{\mathrm{Fi}} R_{c,\mathrm{Fb}}$ is the fraction recycled or retained for burn-up in the fission blanket. Using Eq. (8.28), gives the fissile breeding ratio of the ^{6}Li hybrid as

$$C_{\mathrm{Fi},6} = \eta_o - 1 - \delta_n \tag{8.31}$$

In the limit of negligible neutron losses defining the upper bound, this ^{6}Li hybrid breeding ratio can be written as

$$C_{\mathrm{Fi},6} \simeq \eta_o - 1$$

$$\simeq (\eta_{\mathrm{Fi}} - 1) + (\nu_x - 1)\frac{R_{xn}}{R_{a,\mathrm{Fi}}} + (\nu_{\mathrm{Fb}} - 1)\frac{R_{\mathrm{fi,Fb}}}{R_{a,\mathrm{Fi}}} \tag{8.32}$$

This, of course, possesses some algebraic similarity with the breeding ratio of a fission reactor. The important difference now, however, lies in the fact that η_{Fi} can be much larger since, in the MeV domain we have $\eta_{\mathrm{Fi}} \simeq \nu_{\mathrm{Fi}}$ and ν_{Fi} varies approximately linearly with energy, Eq. (2.1). Further, the fact that cross sections for (n, xn) and fertile fission reactions become significant in the higher energy domain (Figures 2.7 and 2.10) substantially aids the breeding characteristics of this ^{6}Li-hybrid breeder.

8.5. THE ^{7}Li HYBRID

As defined by Figure 8.3b, the ^{7}Li hybrid differs from the ^{6}Li hybrid by the replacement of ^{6}Li by ^{7}Li as fusile–fertile feed material. Then Eq. (8.5) is replaced by Eq. (8.6)

$$^7\mathrm{Li}(n, n'\alpha)\mathrm{T} \tag{8.33}$$

and the rate at which this reaction proceeds in the blanket is given by

$$R_{c,7} = \int_B \int_E \Sigma_{c,7}(E, \mathbf{x})\phi(E, \mathbf{x})dE\, d\mathbf{x} \tag{8.34}$$

The remaining rate equations discussed in the preceding section for the ^{6}Li hybrid are equally applicable to the ^{7}Li hybrid and will be used.

We write, by reference to Eqs. (8.14)–(8.16), or starting from basic particle balance considerations, the relevant rate equations for the ^{7}Li hybrid as

$$\frac{d}{dt}N_T = R_{c,7} - R_{\mathrm{DT}} \tag{8.35}$$

$$\frac{d}{dt}N_{\mathrm{Fi}} = \varepsilon_{\mathrm{Fi}} R_{c,\mathrm{Fb}} - R_{a,\mathrm{Fi}} \tag{8.36}$$

and

$$\frac{d}{dt}N_n = R_{\mathrm{DT}} + (\eta_{\mathrm{Fi}} - 1 - \delta_n)R_{a,\mathrm{Fi}} - R_{c,\mathrm{Fb}} + (\nu_x - 1)R_{xn} + (\nu_{\mathrm{Fb}} - 1)R_{\mathrm{fi,Fb}} \tag{8.37}$$

We note here that, except for the absence of the $R_{c,6}$ term in Eq. (8.37) and the replacement of $R_{c,6}$ by $R_{c,7}$ in Eq. (8.35), this set of rate equations is identical to those for the ^{6}Li hybrid; this feature is due to the (n, n')-type reaction in ^{7}Li which provides for an enhanced neutron economy relative to T breeding in ^{6}Li.

The imposition of steady-state conditions, Eq. (8.17), on the ^{7}Li hybrid rate equations, Eqs. (8.35)–(8.37) can lead to an analysis somewhat similar to that previously developed for the ^{6}Li hybrid though it will yield some different results. From Eqs. (8.35)–(8.37) we find, for steady-state conditions:

$$R_{c,7} = R_{\mathrm{DT}} \tag{8.38}$$

$$\varepsilon_{\mathrm{Fi}} R_{c,\mathrm{Fb}} = R_{a,\mathrm{Fi}} \tag{8.39}$$

$$R_{\mathrm{DT}} + (\eta_{\mathrm{Fi}} - 1 - \delta_n)R_{a,\mathrm{Fi}} = R_{c,\mathrm{Fb}} - (\nu_x - 1)R_{xn} - (\nu_{\mathrm{Fb}} - 1)R_{\mathrm{fi,Fb}} \tag{8.40}$$

We eliminate the breeding reaction $R_{c,\mathrm{Fb}}$ to obtain

$$\frac{R_{\mathrm{DT}}}{R_{a,\mathrm{Fi}}} = \frac{1}{\varepsilon_{\mathrm{Fi}}} + \delta_n - (\eta_0 - 1) \tag{8.41}$$

where we have again made use of the overall neutron yield η_0 per fissile nucleus consumed, Eq. (8.28).

The fusion and fission reaction rates, and therefore their respective powers, are tightly linked by the reactor parameters. One consequence is that the fissile breeding ratio of the ^{7}Li hybrid is determined, in part, by the $R_{\mathrm{DT}}/R_{a,\mathrm{Fi}}$ ratio, as can be seen from Eq. (8.41) using the relationship Eq. (8.30); this also shows that the breeding ratio for a hybrid is given in terms of ε_{Fi}. From Eq. (8.41) we get therefore

$$C_{Fi,7} = \frac{1}{\varepsilon_{Fi}}$$

$$= (\eta_0 - 1) - \delta_n + \frac{R_{DT}}{R_{a,Fi}} \tag{8.42}$$

Comparing this with the ^{6}Li-hybrid breeding ratio, Eq. (8.31), indicates that

$$C_{Fi,6} \lessapprox C_{Fi,7} \tag{8.43}$$

as the neutron spectrum can be expected to be little changed.

The more important result here is that the breeding ratio for the ^{7}Li-hybrid increases with the fusion-to-fission reaction rate ratio and therefore can be made arbitrarily large by increasing the fusion reaction rate in the core relative to the fission reaction rate in the blanket, finally resulting in a fission suppressed symbiont type of system.

Some additional remarks concerning Eq. (8.42) are, however, necessary. We have described the fission/breeding blanket surrounding the fusion core as a subcritical fission reactor driven by the fusion neutron source. As in any driven subcritical assembly, the neutron flux in the blanket should therefore increase in proportion to the source strength, implying

$$\frac{R_{DT}}{R_{a,Fi}} = \text{const} \tag{8.44}$$

In practice, however, the fission rate would be limited by the heat removal capacity from the blanket. For increased R_{DT} the blanket would have to be redesigned by, for example, lowering the fissile fuel enrichment and imposing

$$P_{DTB,t} = \text{const} \tag{8.45}$$

With R_{DT} increasing for $R_{a,Fi}$ kept constant, the 14-MeV source contribution to the neutron spectrum in the blanket becomes more important nd the spectrum becomes harder. This, in turn, tends to increase η_0, nd constitutes a second mechanism which increases the fissile breeding atio with increasing fusion power.

These considerations allude to two vital features of hybrid reactors.)ne is that the breeding ratio can—if desired—be large and, in particular, ery much larger than for a fast breeder reactor. However, this comparison y itself can be misleading because of the importance of the power balance f the hybrid. This latter point becomes evident by noting that for a fusion omponent with an overall electric-to-fusion power multiplication of less han unity, any increase in R_{DT} with $P_{\mathrm{DTB},t}$ constant tends to decrease he net station power output. Even for multiplications equal to or slightly xceeding unity, an increase in R_{DT} results in an increase in the amount f thermal energy to be handled and in the amount of electric power to e recirculated without substantial change of the station power output.

Thus, while at $P_{\mathrm{DTB},t}$ = const the breeding ratio is enhanced with ncreasing fusion power, the net station power is reduced. This feature lludes to the necessity of having to incorporate trade-off considerations etween the power and fissile fuel production of a hybrid. That is, there nay be some choice between having a "fuel factory" or "energy factory."

As a last consideration where we note that for the special case of issile fuel self-sufficiency for the ^{7}Li hybrid, that is $\varepsilon_{\mathrm{Fi}} = 1$, we rewrite Eq. (8.41) to show that the fusion to fission ratio must exactly satisfy the ollowing relation:

$$\frac{R_{\mathrm{DT}}}{R_{a,\mathrm{Fi}}} = 2 + \delta_n - \eta_0 > 0 \tag{8.46}$$

The precise coupling between the fusion and fission power and the fission/breeding reactor blanket parameters thus becomes evident.

8.6. THE Natural-Li HYBRID

A fusion blanket containing lithium—consisting of 7.4% ^{6}Li and 92.6% ^{7}Li—or lithium slightly enriched in ^{6}Li is of more immediate interest. The resultant rate equations will contain neutron-induced reaction in both lithium isotopes and hence contain both $R_{c,6}$ and $R_{c,7}$ terms. The corresponding terms from the previously derived equations could be combined or, alternatively, one might establish the rate equations accordingly as

$$\frac{d}{dt} N_T = -R_{\text{DT}} + R_{c,6} + R_{c,7} \tag{8.47}$$

$$\frac{d}{dt} N_{\text{Fi}} = \varepsilon_{\text{Fi}} R_{c,\text{Fb}} - R_{a,\text{Fi}} \tag{8.48}$$

and

$$\frac{d}{dt} N_n = R_{\text{DT}} + (\eta_{\text{Fi}} - 1 - \delta_n) R_{a,\text{Fi}} - R_{c,\text{Fb}} - R_{c,6} + (\nu_x - 1) R_{xn} + (\nu_{\text{Fb}} - 1) R_{\text{fi,Fb}} \tag{8.49}$$

Examination of this set of equations for the case of steady-state operation shows that the fissile fuel breeding characteristics are in a first approximation in the interval between those of the ^{6}Li and ^{7}Li hybrid:

$$C_{\text{Fi},6} \lessapprox C_{\text{Fi,Nat}} \lessapprox C_{\text{Fi},7} \tag{8.50}$$

Specifically, we find the pertinent expression involving the fusion-to-fission reaction rate ratio to be given by

$$\frac{R_{\text{DT}} - R_{c,6}}{R_{a,\text{Fi}}} + (\eta_0 - 1) - \delta_n - \frac{1}{\varepsilon_{\text{Fi}}} = 0 \tag{8.51}$$

Here, the capture-to-absorption ratio $R_{c,6}/R_{a,\text{Fi}}$ is an important term. However, in view of the feature that the capture cross section in ^{6}Li possesses a magnitude and neutron energy dependence similar to that of fission cross section of fissile nuclei—which becomes evident from a comparison of Figure 2.7 and Figure 2.8—this term cannot be simply expressed except in the limit of zero ^{6}Li concentration; then, however, this would yield the result of the ^{7}Li-hybrid analysis.

8.7. THE T HYBRID

The T hybrid, Figure 8.3c, differs from the previous systems by the total absence of the tritium breeding requirement and therefore of the need to have lithium in the blanket. Here, tritium is assumed to be supplied from external sources such as specially built or modified fission reactors. Alternatively, since all water-cooled reactors—particularly heavy water reactors—unavoidably breed some tritium, and since tritium appears as a fission product in ternary fission, it might be extracted from existing fission reactors.

The reactor representation for the T hybrid, Eq. (8.12), simplifies to

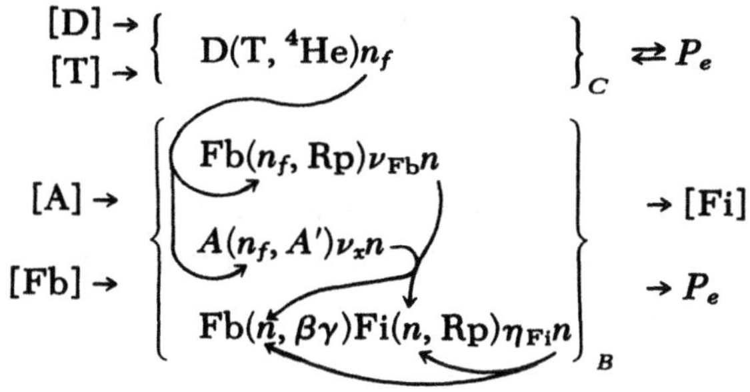

(8.52)

The pertinent rate equations are

$$\frac{d}{dt} N_T = S_T - R_{\rm DT} \tag{8.53}$$

$$\frac{d}{dt} N_{\rm Fi} = \varepsilon_{\rm Fi} R_{c,\rm Fb} - R_{a,\rm Fi} \tag{8.54}$$

and

$$\frac{d}{dt} N_n = R_{\rm DT} + (\eta_0 - 1 - \delta_n) R_{a,\rm Fi} - R_{c,\rm Fb} \tag{8.55}$$

where η_0 again represents the overall neutron yield per fissile atom consumed and includes (n, xn) and fertile fission reactions. Furthermore S_T is the tritium supply rate.

An examination of these equations under steady-state conditions yields results essentially equivalent to the ^{7}Li hybrid with the distinction that, as is evident by Eq. (8.53), the fusion rate is fully determined by the tritium supply rate S_T. The associated fissile destruction rate in the blanket is then

$$R_{a,\mathrm{Fi}} = \left[\frac{S_T}{1/\varepsilon_{\mathrm{Fi}} + 1 + \delta_n - \eta_0}\right] \tag{8.5}$$

In this sense, therefore, the fusion and fission reaction rates are agai tightly coupled and the breeding ratio for this T hybrid, $C_{\mathrm{Fi,T}}$, is given b

$$C_{\mathrm{Fi,T}} = (\eta_0 - 1) - \delta_n + \frac{S_T}{R_{a,\mathrm{Fi}}} \tag{8.5}$$

Note that, by physical and algebraic analogy to the conclusions concernin the breeding ratio for the ^{7}Li hybrid, the tritium supply rate S_T fulfil the same role as R_{DT}.

8.8. TRITIUM TRAJECTORY

The temporal variations of the fusile and fissile fuel requirements c a hybrid reactor need to be taken into account in the design and operatio of a hybrid reactor. As shown in Chapter 5 and Chapter 7 for the cas of a spallation symbiont and D–T symbiont, the nuclear fuel flow charac teristics are functions of several reactor design and operational parameter

In our study of the hybrid fuel flow characteristics, we will find again useful to employ the external stockpile concept. For the case of th D–T hybrid, however, we have a single reactor associated with the tw external stockpiles of tritium and fissile fuel, Figure 8.4. Tritium is bre in the blanket by neutron capture in ^{6}Li and ^{7}Li, separated on sit transferred to the tritium stockpile, and then withdrawn from the stockpi as required to fuel the fusion core. Fissile fuel is similarly bred by neutro capture in the fertile nuclei, separated, transferred, and withdrawn fro the stockpile to fuel the subcritical blanket. Any excess of fissile fuel ma be used to supply external fission reactors.

Each of these two fuel flows will be examined individually usin our previously employed methodology. For the hybrid reactor here, w again choose to use specific breeding reaction rates as this minimizes th need to introduce breeding ratios.

The tritium trajectory calculation for the hybrid is similar to th symbiont of Chapter 7 except for the feature that there is only one blank in which tritium can be bred. Therefore most of the considerations fro Chapter 7 can be applied to the hybrid.

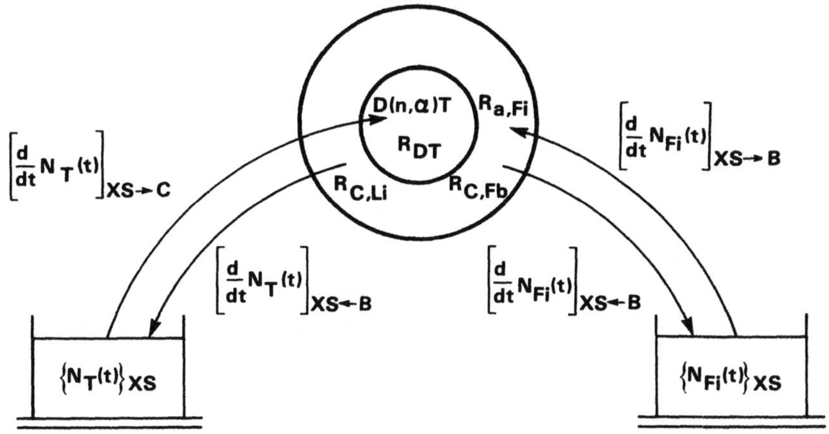

FIGURE 8.4. Fissile and fusile fuel flow between the D–T hybrid reactor and its associated stockpiles.

Referring to Figure 8.4, the differential change in the number of tritium nuclei in the stockpile, $\{N_T(t)\}_{XS}$, is given by an accounting of each of the pertinent processes:

$$\left\{\frac{d}{dt}N_{\mathrm{T}}(t)\right\}_{\mathrm{XS}} = -R_{\mathrm{DT}} - \lambda_{\mathrm{T}}\{N_{\mathrm{T}}(t)\}_{\mathrm{XS}} + \left[\frac{d}{dt}N_{\mathrm{T}}(t)\right]_{\mathrm{XS}\leftarrow B} \tag{8.58}$$

where the { } brackets again refer to system components and the [ ] brackets refer to transfers between the various components. The first term is the tritium supply rate to the core, the next term characterizes the radioactive decay, and the last term identifies the flow rate of the bred tritium from the blanket to the stockpile.

Making the assumptions that the tritium transfer is instantaneous and that the bred tritium has a mean residence time τ_{T} in the blanket allows the last term of Eq. (8.58) to be written as

$$\left[\frac{d}{dt}N_{\mathrm{T}}(t)\right]_{\mathrm{XS}\leftarrow B} = \frac{1}{\tau_{\mathrm{T}}}\{N_{\mathrm{T}}(t)\}_{B} \tag{8.59}$$

The term $\{N_{\mathrm{T}}(t)\}_B$ is the tritium inventory in the blanket of the hybrid. Since it is produced by neutron capture, lost by radioactive decay, and withdrawn from the blanket to the stockpile at a specified rate, we write the differential change in the blanket tritium inventory as

$$\left\{\frac{d}{dt}N_T(t)\right\}_B = R_{c,\text{Li}} - \lambda_T\{N_T(t)\}_B - \frac{1}{\tau_T}\{N_T(t)\}_B \tag{8.60}$$

where $R_{c,\text{Li}} = R_{c,6} + R_{c,7}$ and the meaning of the other terms is self-evident.

Solving Eq. (8.60) for the initial condition $\{N_T(t)\}_B = 0$ for reactor start-up at $t = 0$ and inserting the solution into Eq. (8.59) and, in turn, into Eq. (8.58) yields a differential equation for the tritium inventory in the stockpile in terms of reaction rates and the appropriate reactor parameters

$$\left\{\frac{d}{dt}N_T(t)\right\}_{XS} = -R_{DT} - \lambda_T\{N_T(t)\}_{XS} + \frac{N_{T,B}^{\infty}}{\tau_T}\left\{1 - \exp\left[-t\left(\lambda_T + \frac{1}{\tau_T}\right)\right]\right\} \tag{8.61}$$

where $N_{T,B}^{\infty} = R_{c,\text{Li}}/[(\lambda_T + 1)/\tau_T]$ is the asymptotic blanket inventory. This equation can also be solved subject to the specification of an initial condition for the stockpile inventory

$$\{N_T(t)\}_{XS} = N_{T,XS}^0 - N_{T,C}^0, \qquad t = 0^+ \tag{8.62}$$

where $N_{T,C}^0$ is the initial tritium fuel inventory as required. The temporal variation of the tritium inventory in the external stockpile is found as

$$\{N_T(t)\}_{XS} = \begin{cases} N_{T,XS}^0 \exp(-\lambda_T t), & t \le 0 \\ (N_{T,XS}^0 - N_{T,C}^0)\exp(-\lambda_T t) - \dfrac{R_{DT}}{\lambda_T}[1 - \exp(-\lambda_T t)] & \\ \quad + N_{T,B}^{\infty}\left[\left(\dfrac{1 - \exp(-\lambda_T t)}{\lambda_T \tau_T}\right) - \exp(-\lambda_T t) \right. & \\ \quad \left. \times [1 - \exp(-t/\tau_T)\right], & t \ge 0 \end{cases} \tag{8.63}$$

We display some characteristic trajectories in Figure 8.5.

As suggested in Chapter 7 for the symbiont, this stockpile trajectory can be used to identify some important features of the tritium fuel flow associated with a given hybrid reactor and its tritium management. Com-

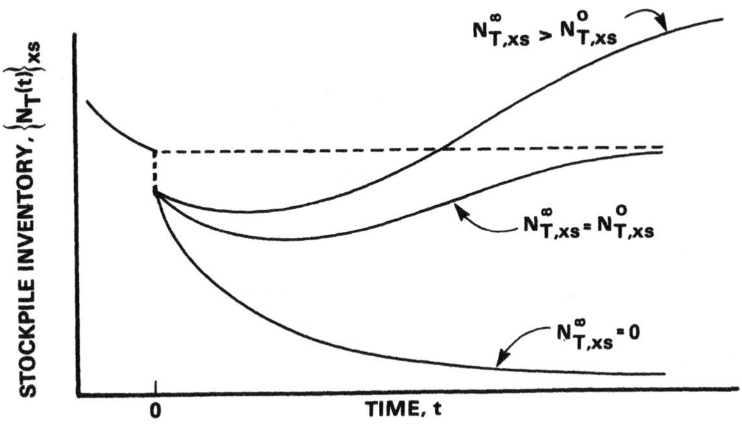

FIGURE 8.5. Trajectories of the tritium inventory in the tritium stockpile.

non considerations would involve the asymptotic stockpile inventory, ninimum tritium requirement for start-up and operation, and others. In general the hybrid tritium trajectory resembles very much the tritium rajectory discussed for the symbiont. In both cases, the energy system should be rendered self-sufficient with respect to tritium. In the hybrid case all tritium has to be bred in the blanket surrounding the fusion core, which may result in a severe penalty for the fissile fuel breeding capacity of this same blanket.

8.9. FISSILE TRAJECTORY

The time variation of the fissile fuel inventory in the fissile fuel stockpile can be established in a manner similar to that developed for tritium. Radioactive decay of the bred fissile fuel will, however, be neglected. Initial blanket fissile fuel loading is likely required to ensure uniformity of power production prior to attaining equilibrium. Proceeding thus yields the following expression for the fissile fuel inventory in the fissile stockpile as a function of time:

$$\{N_{\mathrm{Fi}}(t)\}_{\mathrm{XS}} = N^{0}_{\mathrm{Fi,XS}} - N^{0}_{\mathrm{Fi},B} + (R_{c,\mathrm{Fb}} - R_{a,\mathrm{Fi}})t - [\tau_{\mathrm{Fi}} R_{c,\mathrm{Fb}} - N^{0}_{\mathrm{Fi},B}][1 - \exp(-t/\tau_{\mathrm{Fi}})] \qquad (8.64)$$

where τ_{Fi} is the mean residence time of the fissile fuel in the blanket. We display a typical trajectory of this fissile stockpile inventory $\{N_{\mathrm{Fi}}(t)\}_{\mathrm{XS}}$

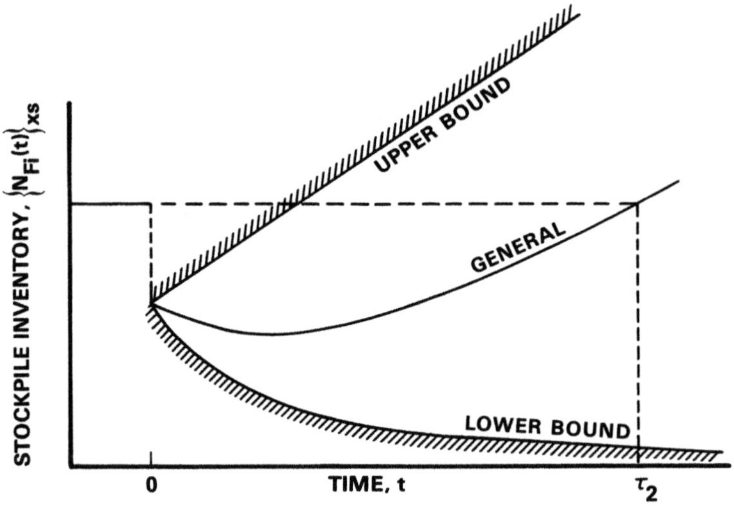

FIGURE 8.6. Trajectory bounds of the fissile fuel in the fissile stockpile.

located between two relevant bounds in Figure 8.6. The upper bound in this figure is associated with the case of $R_{c,\mathrm{Fb}} \gg R_{a,\mathrm{Fi}}$ and $\tau_{\mathrm{Fi}} \to 0$. Such conditions would make the hybrid a super breeder of fissile fuel.

The lower bound is associated with an asymptotic fissile self-sufficient state for which $R_{c,\mathrm{Fb}} = R_{a,\mathrm{Fi}}$ and $\tau_{\mathrm{Fi}} > 0$. In a sense, this hybrid could be viewed as one possessing a fissile fuel breeding ratio of unity, although a decreasing amount of fissile fuel would have to be supplied as the asymptotic steady state is approached or, in our model, as the blanket inventory is building up to its asymptotic value.

A more typical fissile fuel trajectory will be one located between these bounds. We have indicated here the time coordinate τ_2 as corresponding to that time when the fissile stockpile attains its original level. This means that at $t = \tau_2$ a new identical hybrid reactor could be supplied with sufficient fissile fuel to sustain the same fissile fuel flow pattern. This τ_2 could therefore be interpreted as a fissile fuel replenishment time or as a hybrid doubling time. We consider in the following an assessment of this time τ_2 as an example of how point parameters can be extracted from the fuel trajectory formulation.

At $t = \tau_2$ we have

$$\{N_{\mathrm{Fi}}(\tau_2)\}_{\mathrm{XS}} = N^0_{\mathrm{Fi,XS}} \tag{8.65}$$

and with the aid of Eq. (8.64) obtain an equation for τ_2:

$$N^0_{\mathrm{Fi},B} = (R_{c,\mathrm{Fb}} - R_{a,\mathrm{Fi}})\tau_2 - (\tau_{\mathrm{Fi}} R_{c,\mathrm{Fb}} - N^0_{\mathrm{Fi},B})[1 - \exp(-\tau_2/\tau_{\mathrm{Fi}})] \tag{8.66}$$

Restricting ourselves to the case for which $\tau_{\text{Fi}} \ll \tau_2$ leads immediately to

$$\tau_2 \simeq \frac{N^0_{\text{Fi},B} + \tau_{\text{Fi}} R_{c,\text{Fb}}}{R_{c,\text{Fb}} - R_{a,\text{Fi}}}, \qquad R_{c,\text{Fb}} > R_{a,\text{Fi}} \tag{8.67}$$

To cast this analysis into a form somewhat resembling a fission reactor analysis—which seems fully justified since the blanket of a hybrid is expected to possess many of the properties of a fission reactor—we introduce a blanket fissile fuel breeding ratio, $C_{\text{Fi},B}$:

$$C_{\text{Fi},B} = R_{c,\text{Fb}}/R_{a,\text{Fi}} \tag{8.68}$$

With this definition we write Eq. (8.67) in the more compact form

$$\tau_2 = \frac{N^0_{\text{Fi},B}/R_{a,\text{Fi}} + \tau_{\text{Fi}} C_{\text{Fi},B}}{C_{\text{Fi},B} - 1} \tag{8.69}$$

The ratio $N^0_{\text{Fi},B}/R_{a,\text{Fi}}$ has dimensions of time and characterizes the initial fissile fuel load in units of the annual gross fuel consumption. It will clearly vary from one blanket to another depending upon design and operational details.

8.10. HYBRID POWER BALANCE

The electrical power output of the DT-based hybrid reactor, Figure 8.2, is given by

$$P_{o,e} = \eta_{B,g} P_{B,t} - P_{\text{ci},e} - P_{\text{oc},e} \tag{8.70}$$

where $P_{B,t}$ is the total thermal power deposited and generated in the blanket and $\eta_{B,g}$ is the associated gross thermal-to-electrical conversion efficiency for transforming this thermal power into electrical; $P_{\text{ci},e}$ is the circulating electrical power required to sustain the fusion reactions and $P_{\text{oc},e}$ is the electrical power required for station operation and control. No direct charged particle conversion is taken to occur.

We take the $P_{\text{oc},e}$ component to be a fraction of the thermal blanket power

$$P_{\text{oc},e} = \varepsilon_{B,\text{oc}} P_{B,t} \tag{8.71}$$

and write Eq. (8.70) more compactly as

$$P_{o,e} = \eta_B P_{B,t} - P_{\text{ci},e} \tag{8.72}$$

where η_B is the net blanket power conversion efficiency defined here as

$$\eta_B = \eta_{B,g} - \varepsilon_{B,\text{oc}} \tag{8.73}$$

Incorporating the fusion component is done in a manner similar to that employed in Chapter 6. We define a hybrid power multiplication M_{hy} by

$$M_{\text{hy}} = \frac{\eta_B P_{B,t}}{P_{\text{ci},e}} \tag{8.74}$$

where now the fission processes may be a dominant contribution to the reactor power multiplication. Indeed, writing the blanket thermal power as a sum of the various thermal power components gives

$$P_{B,t} = P_n + P_r + \sum_i P_{B,i} + P_{\text{fi},t} \tag{8.75}$$

where the first three terms are used as defined in Section 6.4, and Eq. (6.28). The last term is fission power and, for a hybrid reactor, will likely be the largest contributor to the power produced.

With Eq. (8.74), the station power may therefore be reduced to

$$P_{o,e} = \eta_B(1 - 1/M_{\text{hy}})P_{B,t} \tag{8.76}$$

A blanket power multiplication may be defined for the fission blanket in a manner similar to that employed for the symbiont, Eq. (6.29). For the system we obtain

$$\begin{aligned} M_{\text{hy},B} &= \frac{P_{B,t}}{P_n + P_r} \\ &= 1 + \frac{\sum_i P_{B,i}}{P_n + P_r} + \frac{P_{\text{fi},t}}{P_n + P_r} \\ &= M_B + \frac{P_{\text{fi},t}}{P_n + P_r} \end{aligned} \tag{8.77}$$

where M_B is the blanket multiplication of a nonfission blanket, Eq. (6.30).

The fission blanket power multiplication, $M_{hy,B}$, may be used as a convenient parameter and incorporated into the station power expression. We write Eq. (8.76) therefore as

$$P_{o,e} = \eta_B(1 - 1/M_{hy})(P_n + P_r)M_{hy,B} \tag{8.78}$$

To have the fusion power appear explicitly, we substitute, as in Eqs. (6.43) and (6.45)

$$P_n + P_r = (0.8 + \gamma_r)P_{DT} \tag{8.79}$$

and obtain

$$P_{o,e} = \eta_B(1 - 1/M_{hy})(0.8 + \gamma_r)P_{DT}M_{hy,B} \tag{8.80}$$

In the limit of having the fission power contribution be dominant we then approximate Eq. (8.77) by

$$M_{hy,B} \sim \frac{P_{fi,t}}{P_n + P_r} \tag{8.81}$$

and

$$P_{o,e} \sim \eta_B(1 - 1/M_{hy})P_{fi,t} \tag{8.82}$$

The hybrid reactor power multiplication, however, continues to provide the connection with the efficiency of fusion energy production.

PROBLEMS

8.1. Formulate an "equivalent" fertile–fusile fuel concept for the treatment of ^6Li and ^7Li.

8.2. Confirm the validity of Eqs. (8.50) and (8.51).

8.3. Examine $\{N_T(t)\}_{XS}$, Eq. (8.63), for its tritium breeding doubling time to supply other T-hybrid reactors.

8.4. In Chapters 6 and 8 different fuel flow descriptions are used: in Chapter 6 it was based on conversion ratios, and, alternatively, in Chapter 8 we used the neutron balance equation. Reformulate the basic fuel flow equations in Chapter 8 with the methodology of Chapter 6.

8.5. Plot and interpret the tritium and the fissile trajectory of a fusion–fission hybrid using standard parameter values for the system components.

8.6. Calculate and plot (i) the net hybrid power output and; (ii) the recirculating power of a fusion–fission hybrid as a function of the hybrid power multiplication parameter M_{hy} for various blanket multiplication factors. Use standard parameter values for the system components.

Part IV

THE BROADER PERSPECTIVE

Chapter Nine

THE RECEDING HORIZON

In the preceding chapters we focused on nuclear energy technologies haracterized by heavy-element fission, D–T fusion, spallation processes, nd central station power plants. Our interest in this chapter is to look beyond these systems and identify other fuels as well as other synergetic nuclear energy systems concepts.

9.1. D–D FUSION

While the first use of fusion energy will by all indications be based on the fusion of deuterium and tritium, it is envisaged that subsequent usion reactors will be fueled by deuterium only. Though plasma conditions then become more demanding, the feature that these fusion fuels are not resource limited and do not require fuel breeding renders them most appealing.

In Chapter 2 we introduced three distinct operational modes associated with deuterium fueled fusion: PURE-D, SCAT-D, and CAT-D. These hree modes were characterized by the varying degree with which ion control was maintained and the progeny ions fuse with the primary fuels. We show in Figure 9.1 the reactivities for primary ion fusion, D + D, as well as the reactivities for reactions involving the progeny ions with the primary ions, D + T and D + ^{3}He; in the same figure we also display the reactivities involving the progeny ions among themselves, T + T, He + ^{3}He, and T + ^{3}He. As shown, fusion reactions involving these

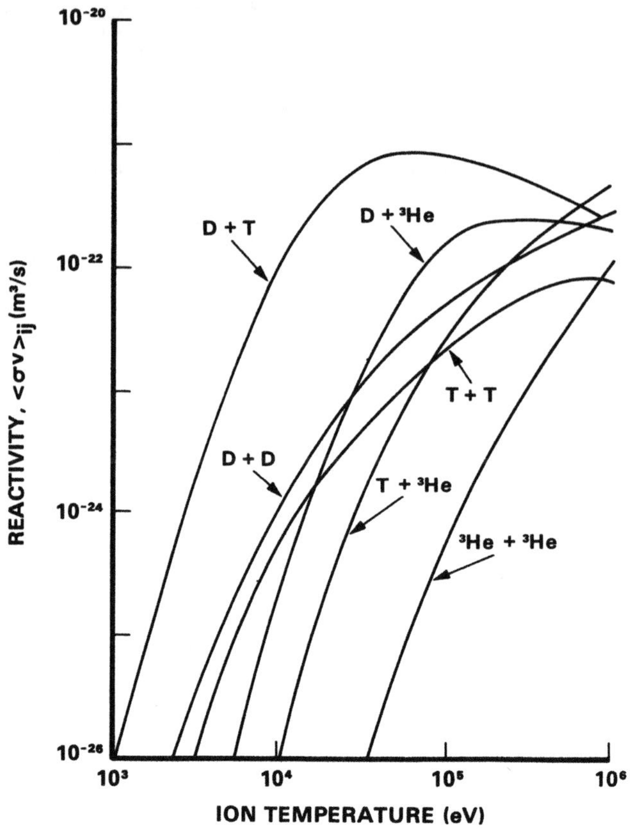

FIGURE 9.1. Temperature dependence of reactivities of selected fusion reactions.

higher-order products occur with increasing likelihood at higher plasma temperatures.

We write an arbitrary fusion reaction rate as

$$R_{ij} = \frac{n_i n_j}{1 + \delta_{ij}} \langle \sigma v \rangle_{ij} V \tag{9.1}$$

with V as the reaction volume. However, with the D–D fusion channels represented by

$$\mathrm{D} + \mathrm{D} \begin{cases} \nearrow \mathrm{T} + p, & R_{1,\mathrm{DD}} \text{ (channel No.1)} \\ \searrow {}^3\mathrm{He} + n, & R_{2,\mathrm{DD}} \text{ (channel No. 2)} \end{cases} \tag{9.2}$$

we specify

$$R_{k,\mathrm{DD}} = \frac{n_D^2}{2}\langle\sigma v\rangle_{k,\mathrm{DD}} V, \qquad k = 1, 2 \tag{9.3}$$

with the index k here used to distinguish the two channel reactions. The associated reactivities $\langle\sigma v\rangle_{k,\mathrm{DD}}$ must be distinguished from the reactivity for D + D fusion $\langle\sigma v\rangle_{\mathrm{DD}}$; the relationship between them can be characterized by a pair of branching ratios, $\varepsilon_{k,\mathrm{DD}}$:

$$\langle\sigma v\rangle_{k,\mathrm{DD}} = \varepsilon_{k,\mathrm{DD}}\langle\sigma v\rangle_{\mathrm{DD}}, \qquad k = 1, 2 \tag{9.4}$$

It is generally adequate to take

$$\varepsilon_{1,\mathrm{DD}} = \varepsilon_{2,\mathrm{DD}} \tag{9.5}$$

which asserts that the D–D channels occur with an equally likely probability. Precise measurements have shown that there does exist a small difference between $\langle\sigma v\rangle_{1,\mathrm{DD}}$ and $\langle\sigma v\rangle_{2,\mathrm{DD}}$ which varies with temperature.

The reactions possible in a D–D fueled fusion reactor may hence be represented by the coupled reaction system displayed in Table 9.1. Here we have introduced reaction branching ratios ε_{ij} to characterize the linkages of the various reactions; in order to distinguish this general case of D–D fusion from the PURE-D, SCAT-D, and CAT-D modes, we choose to designate it as the extended, EXT-D mode. Examination of Table 9.1 suggests that at steady state the following conditions among the reaction rates hold:

$$\varepsilon_{\mathrm{TD}} R_{1,\mathrm{DD}} = R_{\mathrm{TD}} \tag{9.6a}$$

$$\varepsilon_{\mathrm{TT}} R_{1,\mathrm{DD}} = R_{\mathrm{TT}} \tag{9.6b}$$

$$\varepsilon_{\mathrm{THe}} R_{1,\mathrm{DD}} = R_{\mathrm{THe}} \tag{9.6c}$$

$$\varepsilon_{\mathrm{HeT}} R_{2,\mathrm{DD}} = R_{\mathrm{HeT}} \qquad (\text{where } R_{\mathrm{THe}} = R_{\mathrm{HeT}}) \tag{9.6d}$$

$$\varepsilon_{\mathrm{HeD}} R_{2,\mathrm{DD}} = R_{\mathrm{HeD}} \tag{9.6e}$$

$$\varepsilon_{\mathrm{HeHe}} R_{2,\mathrm{DD}} = R_{\mathrm{HeHe}} \tag{9.6f}$$

TABLE 9.1
Details of the EXT-D Mode Cycle Illustrating Coupling and Associated Branching Ratios, ε_{ij}, Reaction Energies $U_{k,ij}$, and Reaction Rates $R_{k,ij}$[a]

Reaction system	$U_{k,ij}$	$R_{k,ij}$
$D + D \rightarrow T(1.01) + p(3.02)$	$U_{1,DD}$	$R_{1,DD}$
ε_{TD}		
$D + T \rightarrow n(14.07) + \alpha(3.52)$	U_{TD}	R_{TD}
ε_{TT}		
$T + T \rightarrow n(5.03) + n(5.03) + \alpha(1.26)$	U_{TT}	R_{TT}
ε_{THe}		
$T + He \rightarrow p(5.69) + \alpha(1.28) + n(5.12)$	U_{THe}	R_{THe}
ε_{HeT}		
$D + D \rightarrow He(0.82) + n(2.45)$	$U_{2,DD}$	$R_{2,DD}$
ε_{HeD}		
$D + He \rightarrow \alpha(3.67) + p(14.68)$	U_{DHe}	R_{DHe}
ε_{HeHe}		
$He + He \rightarrow p(5.7) + p(5.7) + \alpha(1.43)$	U_{HeHe}	R_{HeHe}

[a] For reasons of notational simplicity we use He = ^{3}He and α = ^{4}He. The T + He reaction is known to lead to three channels, two of which produce $p + n + \alpha$ and one yielding $D + \alpha$. Here we suppress the $D + \alpha$ channel and use weighted energy assignments for n, p, and α according to their channel probabilities.

where the progeny reaction ratios depend on various plasma parameters and reactor operations.

Particle balance considerations impose specific conditions on these progeny ratios. For the steady state ideal loss-free case we use Table 9.1 to obtain the following constraints

$$\varepsilon_{TD} + 2\varepsilon_{TT} + \varepsilon_{THe} = 1 \tag{9.7a}$$

$$\varepsilon_{HeT} + \varepsilon_{HeD} + 2\varepsilon_{HeHe} = 1 \tag{9.7b}$$

Particle production rates can be written for any of the progeny particles in the form of

$$\frac{dN_m}{dt} = \sum_i \sum_j R_{k_j i_j} \tag{9.8}$$

where the summation is taken over those reaction rates which contribute—positively or negatively—to the particle inventory N_m; with the channel index k taken as appropriate. Similarly, the nuclear powers associated with a given particle are given by

$$P_m = \sum_i \sum_j U_{k,ij} R_{k,ij} \tag{9.9}$$

where $U_{k,ij}$ is the kinetic energy carried by the particle emerging from the pertinent reactions.

In the above, we have indicated only that the branching ratios are positive and satisfy certain additive conditions. Explicit expressions can be derived as a function of the reactivities $\langle \sigma v \rangle_{ij}$ and hence as a function of temperature, or more generally, as a function of the ion energy distribution.

As a final observation, we point to the feature that the previously employed D–D burn modes, that is PURE-D, SCAT-D, and CAT-D, can be recovered from this general EXT-D formulation by a specified imposition on three branching ratios as follows:

$$\text{PURE-D:} \quad \varepsilon_{ij} = 0 \qquad \text{all } i \text{ and } j \tag{9.10a}$$

$$\text{SCAT-D:} \quad \varepsilon_{\text{TD}} = 1$$

$$\varepsilon_{\text{TT}} = \varepsilon_{\text{THe}} = \varepsilon_{\text{HeT}} = \varepsilon_{\text{HeD}} = \varepsilon_{\text{HeHe}} = 0 \tag{9.10b}$$

$$\text{CAT-D} \quad \varepsilon_{\text{TD}} = \varepsilon_{\text{HeD}} = 1$$

$$\varepsilon_{\text{TT}} = \varepsilon_{\text{HeHe}} = \varepsilon_{\text{HeT}} = \varepsilon_{\text{THe}} = 0 \tag{9.10c}$$

9.2. ADVANCED FUSION FUELS

A convenient sequence characterizing fusion fuels and fusion reactions can be represented by the following:

I. Conventional

$$\text{D–T:} \qquad \text{D} + \text{T} \rightarrow n + \alpha + 17.6\ \text{MeV} \tag{9.11}$$

II. Deuterium based

$$\text{PURE-D:}\quad D + D \rightarrow \begin{cases} T + p + 4.0\ \text{MeV} \\ {}^{3}\text{He} + n + 3.3\ \text{MeV} \end{cases} \qquad (9.12)$$

SCAT-D: D + D with T burned *in situ* and ^{3}He extracted (24.9 MeV cycle reaction energy) (9.13)

CAT-D: D + D with T and ^{3}He burned *in situ* (43.2 MeV cycle reaction energy) (9.14)

EXT-D: D + D with concurrent burn, to a varying extent, of all progeny products (9.15)

$$\text{D–}^{3}\text{H:}\quad D + {}^{3}\text{He} \rightarrow \alpha + p + 18.3\ \text{MeV} \qquad (9.16)$$

$$\text{D–}^{6}\text{Li:}\quad D + {}^{6}\text{Li} \rightarrow \begin{cases} {}^{7}\text{Be} + n + 3.4\ \text{MeV} \\ {}^{7}\text{Li} + p + 5.0\ \text{MeV} \\ p + \alpha + T + 2.6\ \text{MeV} \\ 2\alpha + 22.3\ \text{MeV} \\ {}^{3}\text{He} + \alpha + n + 1.8\ \text{MeV} \end{cases} \qquad (9.17)$$

$$\text{D–}^{7}\text{Li:}\quad D + {}^{7}\text{Li} \rightarrow n + 2\alpha + 15.1\ \text{MeV} \qquad (9.18)$$

$$\text{D–}^{7}\text{Be:}\quad D + {}^{7}\text{Be} \rightarrow p + 2\alpha + 16.8\ \text{MeV} \qquad (9.19)$$

etc.

III. Proton based

$$p\text{–}^{6}\text{Li:}\quad p + {}^{6}\text{Li} \rightarrow \alpha + {}^{3}\text{He} + 4.0\ \text{MeV} \qquad (9.20)$$

$$p\text{–}^{9}\text{Be}: \quad p + {}^{9}\text{Be} \rightarrow \begin{cases} \text{D} + 2\alpha + 0.7\ \text{MeV} \\ {}^{6}\text{Li} + \alpha + 2.1\ \text{MeV} \end{cases} \tag{9.21}$$

$$p\text{–}^{11}\text{B}: \quad p + {}^{11}\text{B} \rightarrow 3\alpha + 8.7\ \text{MeV} \tag{9.22}$$

etc.

IV. Various

$$\text{T–T}: \quad \text{T} + \text{T} \rightarrow \alpha + 2n + 11.3\ \text{MeV} \tag{9.23}$$

$$^{3}\text{He–}^{3}\text{He}: \quad {}^{3}\text{He} + {}^{3}\text{He} \rightarrow 2p + \alpha + 12.8\ \text{MeV} \tag{9.24}$$

$$\text{T–}^{3}\text{He}: \quad \text{T} + {}^{3}\text{He} \rightarrow \begin{cases} \text{D} + \alpha + 14.3\ \text{MeV} \\ p + \alpha + n + 12.1\ \text{MeV} \end{cases} \tag{9.25}$$

$$^{6}\text{Li-}^{6}\text{Li}: \quad {}^{6}\text{Li} + {}^{6}\text{Li} \rightarrow \begin{cases} p + {}^{11}\text{B} + 12.2\ \text{MeV} \\ \alpha + n + {}^{7}\text{Be} + 1.9\ \text{MeV} \end{cases} \tag{9.26}$$

Numerous channels in this last expression are not listed but suggested by the notation.

The above listings of fusion reactions also indicate the extent of neutron production. Since neutrons are the dominant particles to induce radioactivity in the structural materials, some fusion reactions may be termed "cleaner" than others. The increasing fraction of charged particles will evidently render it desirable to convert the charged particle energy into electricity by direct conversion. Based on this criterion of minimum neutron production, it is evident that the $p\text{–}^{11}\text{B}$ reaction possesses special appeal as it represents a particularly attractive neutron-free reaction with alpha particles as the only reaction products.

The existence of a set of reaction channels and the potential for fusion involving reaction products with the primary fuel provides conditions which could lead to particular kinds of fusion chains. That is, some fusion products may serve the function of a chain carrier much as neutrons serve the function of maintaining a chain reaction in a fission reactor. We illustrate an example of a fusion chain starting with $p\text{–}^{6}\text{Li}$ fusion. In terms of the reaction linkage notation developed in Chapter 3, we consider the following reaction system:

$$\begin{aligned} &p + {}^{6}\mathrm{Li} \rightarrow {}^{3}\mathrm{He} + \alpha \\ &{}^{6}\mathrm{Li} + {}^{3}\mathrm{He} \rightarrow p + \alpha_f + \alpha \\ &{}^{6}\mathrm{Li} + \alpha_f \rightarrow p + {}^{9}\mathrm{Be} \end{aligned} \tag{9.27}$$

Here, the alpha particle α_f represents the sufficiently energetic ion allowing the subsequent reaction to occur. In this expression we have identified the appearance of two progeny protons which could close the cycle to form a fusion chain. While, in principle, only one proton is necessary to maintain a chainlike process, we indicate here two features which adversely affect the closure of this cycle. First, in the second line of Eq. (9.27), not all of the several channels leads to proton production, and secondly, only one of the two alpha particle reaction products possesses a sufficiently high initial energy to fuse with ${}^{6}\mathrm{Li}$.

In view of the diversity of fusion reactions listed above, one may identify several conceivable fusion chains. The question of the extent to which fusion particle excursion chain processes can occur similar to fission excursion chains is not simply answerable. A point to note here is that the increasing density of a chain carrier would eventually increase the fusion rate with itself as well as with other particles and therefore tend to suppress the tendency for intrinsic self-propagation.

9.3. FUSION SATELLITE REACTORS

The various nuclear energy system concepts referred to in the preceding chapters are generally perceived as large central station power plants. In contrast, the concept of small intrinsically safe compact nuclear reactors regionally distributed to serve small load centers possesses considerable appeal for selected applications. We indicate here how a system of small satellite fusion reactors could be linked to a large central parent reactor.

We consider a central station D–D-fueled fusion reactor operating on the SCAT-D mode. The ${}^{3}\mathrm{He}$ reaction product would be extracted to serve as fuel for small decentralized fusion reactors optimized for the D–${}^{3}\mathrm{He}$ fusion reaction. These satellites might be radiologically relatively clean since the products from this reaction are only charged particles, Eq. (9.16). Furthermore, these charged particles could be converted into electricity

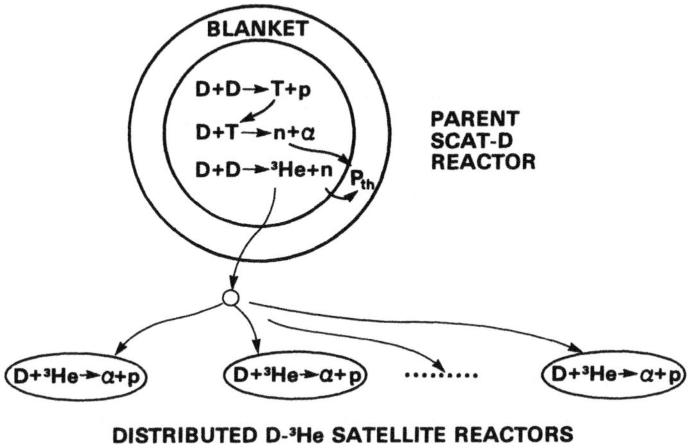

FIGURE 9.2. Schematic depiction of a D–^{3}He parent–satellite reactor system concept.

by direct conversion which, in principle, is possible at a higher efficiency than in the thermodynamic conversion cycle. Additionally, then, the parent reactor could also be equipped with a fissile blanket providing therefore for the production of additional energy and, potentially, fissile fuel. Figure 9.2 provides a schematic representation of this fusion parent–satellite system.

The reaction equations and reaction rates for the SCAT-D parent reactor are

$$\begin{aligned} &D + D \rightarrow T + p, && R_{1,DD} \\ &D + T \rightarrow n + \alpha, && R_{DT} \\ &D + D \rightarrow {}^3He + n, && R_{2,DD} \end{aligned} \tag{9.28}$$

and for the satellites

$$D + {}^3He \rightarrow p + \alpha, \qquad R_{DHe} \tag{9.29}$$

with residual neutron production via D + D reactions. Thus, deuterium is the only primary fuel supplied and stable ^{3}He the only fuel to be transported. If we include fissile breeding and p–α plasma heating, the isotope balance of the entire parent–satellite system could be represented by

$$\begin{array}{lll} 5D + \overbrace{p + \alpha}^{\longrightarrow \text{ plasma heating}} + {}^3He + 2n & & \Big\} \text{Parent core} \\ \quad n + Fb \rightarrow Fi & & \} \text{Parent blanket} \qquad (9.30) \\ {}^3He + D \rightarrow p + \alpha & & \} \text{Satellites} \end{array}$$

For the SCAT-D parent reactor, T is consumed at its rate of production accomplished by the establishment of an equilibrium tritium inventory in the parent fusion core. The ratio of tritium-to-deuterium densities in the core is then found from the imposed equivalence of the reaction rate

$$R_{1,DD} = R_{DT} \tag{9.31}$$

which implies

$$\frac{n_D^2}{2}\langle \sigma v \rangle_{1,DD} = n_D n_T \langle \sigma v \rangle_{DT} \tag{9.32}$$

with the ratio n_T/n_D then given by

$$\frac{n_T}{n_D} = \frac{1}{2}\frac{\langle \sigma v \rangle_{1,DD}}{\langle \sigma v \rangle_{DT}}$$

$$\simeq \frac{1}{4}\frac{\langle \sigma v \rangle_{DD}}{\langle \sigma v \rangle_{DT}} \tag{9.33}$$

Here, use of Eq. (9.5) with $\varepsilon_{k,DD} \simeq 1/2$ has been made. We display this density ratio in Figure 9.3.

The production rate of ^3He as a fuel for the satellites is given by

$$\left\{\frac{d}{dt} N_{He}\right\}_{PAR} = R_{2,DD} \tag{9.34}$$

where the { } brackets refer to the core of the parent SCAT-D reactor PAR. In a steady-state ^3He-self-sufficient operation mode, the production rate of ^3He in the parent core will exactly equal the ^3He destruction rate

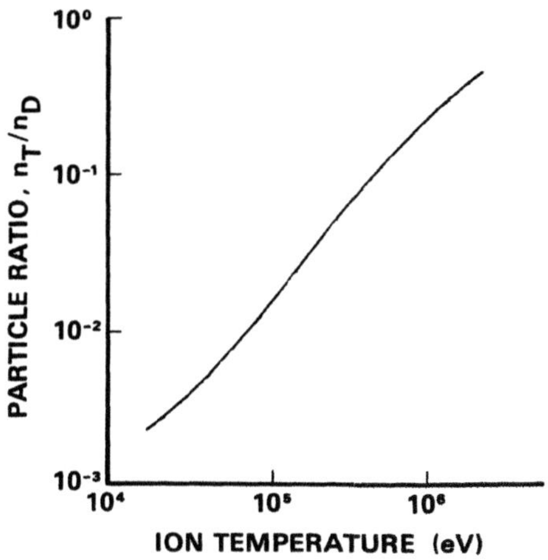

FIGURE 9.3. Equilibrium ion inventories of SCAT D–D plasmas as a function of plasma ion temperature.

in the satellites:

$$\left\{\frac{d}{dt}N_{\mathrm{He}}\right\}_{\mathrm{PAR}} = \left\{-\frac{d}{dt}N_{\mathrm{He}}\right\}_{\mathrm{SAT}} \tag{9.35}$$

This equation set may now be used to determine relevant power characteristics.

9.4. FUSION PARENT–SATELLITE POWER

The electrical output of the SCAT-D parent reactor may be represented by

$$\{P_e\}_{\mathrm{PAR}} = \{\eta_{\mathrm{th}}P_{\mathrm{th}} + \eta_{\mathrm{cp}}P_{\mathrm{cp}} - P_{\mathrm{ci},e}\}_{\mathrm{PAR}} \tag{9.36}$$

Here P_{th} is the thermal energy generated in the parent blanket attributed mainly to the fusion neutron induced processes, Figure 9.2, and, in our notation, given by

$$P_{\mathrm{th}} = (U_{\mathrm{DT}(n)}R_{\mathrm{DT}} + U_{2,\mathrm{DD}}\mathrm{R}_{2,\mathrm{DD}})M_B \tag{9.37}$$

where M_B is the energy multiplication parameter of the blanket. The term P_{cp} in Eq. (9.36) is the fusion core power extracted by direct conversion of charged particles. The factors, η_{th} and η_{cp} are the corresponding net conversion efficiencies.

The net electricity produced by all the satellite reactors is given by

$$\{P_e\}_{\text{SAT}} = \sum_i \{\eta_{\text{cp}} P_{\text{cp}} - P_{\text{ci},e}\}_{\text{SAT},i} \tag{9.38}$$

with i referring to the ith satellite.

The total electrical power from the entire parent–satellite system, including the energy value of the fissile fuel bred in the parent reactor blanket, is given by

$$\begin{aligned} P_{o,e} &= \{P_e\}_{\text{PAR}} + \{P_e\}_{\text{SAT}} \\ &= \{\eta_{\text{th}} P_{\text{th}} + \eta_{\text{cp}} P_{\text{cp}} - P_{\text{ci},e}\}_{\text{PAR}} + \sum_i \{\eta_{\text{cp}} P_{\text{cp}} - P_{\text{ci},e}\}_{\text{SAT},i} \end{aligned} \tag{9.39}$$

The role of the station electric power multiplication (Appendix D) in affecting the total station power output may be assessed by

$$\{M_e\}_{\text{PAR}} = \left\{\frac{\eta_{\text{th}} P_{\text{th}} + \eta_{\text{cp}} P_{\text{cp}}}{P_{\text{ci},e}}\right\}_{\text{PAR}} \tag{9.40}$$

and

$$\{M_e\}_{\text{SAT},i} = \left\{\frac{\eta_{\text{cp}} P_{\text{cp}}}{P_{\text{ci},e}}\right\}_{\text{SAT},i} \tag{9.41}$$

with the total system power therefore represented by

$$\begin{aligned} P_{o,e} &= \{(\eta_{\text{th}} P_{\text{th}} + \eta_{\text{cp}} P_{\text{cp}})(1 - 1/M_e)\}_{\text{PAR}} \\ &\quad + \sum_i \{(\eta_{\text{cp}} P_{\text{cp}})(1 - 1/M_e)\}_{\text{SAT},i} \end{aligned} \tag{9.42}$$

Evidently, some, but not all, power multiplication parameter could indeed be less than unity.

In contrast to the symbiosis of numerous small D–^{3}He satellite reactors coupled to a central SCAT-D fusion reactor, one may conceive of a system of many small fission reactors coupled to one central spallation accelerator facility. The principal function of the small satellites would be to provide low-temperature heat for localized space heating while the spallation accelerator breeds the required fuel and rejuvenates spent fuel for these decentralized regionally distributed nuclear heat batteries. Figure 9.4 provides a schematic representation of this spallation-fission satellite concept we have in mind.

The spallation accelerator of interest here has been discussed in Chapter 4. We take it to be equipped with an appropriate blanket surrounding the target which, by suitable design, contains positions with neutron spectra optimized for *in situ* fuel rejuvenation as discussed in Section 3.6; additionally, fissile fuel breeding should occur in the blanket and selected fission product and/or actinide destruction by transmutation should take place either in the blanket or in the target.

The satellite fission reactors are assumed to be low power (<2 MW$_t$), low temperature (<100°C), slightly enriched (<5%), tank-type water-cooled reactors designed to provide hot water for direct space heating purposes. The absence of steam generators and turbines, but not heat exchangers, thus reduces cost and the low temperature domain and low

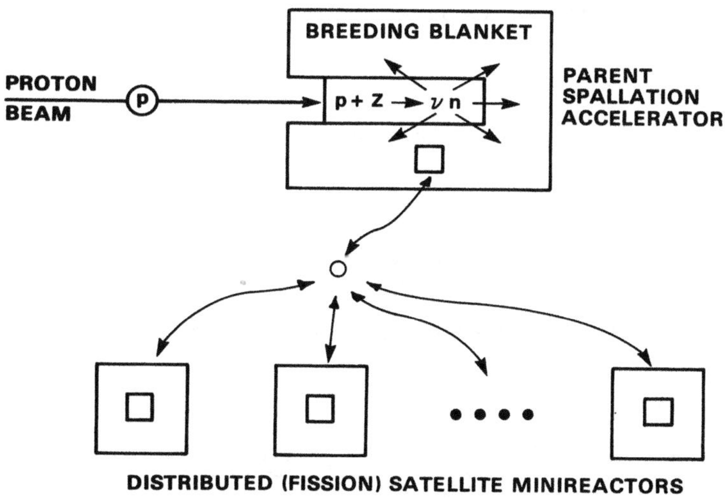

FIGURE 9.4. Graphical depiction of a "parent" accelerator supplying a large number of small fission "satellite" minireactors with fuel; the double arrows are to suggest that spent fuel could be rejuvenated in the accelerator blanket for repeated use.

power level provide substantial safety features sufficient for possible system operation without the full-time attendance of a licensed operator, thus reducing cost further.

An indication of some dominant system characteristics can be obtained from the following simplified analysis based on a fully closed fuel cycle. We consider the total thermal power of the various satellites and the power required to produce the proton beam to be the dominant power components. The power associated with operations, control, fuel fabrication, blanket heat, etc., is taken to be a relatively minor contribution. The system thermal power output is therefore

$$P_{o,t} = \sum_{i} \{P_{\mathrm{fi},t}\}_{\mathrm{SAT},i} - \{P_{p,t}\}_{\mathrm{ACC}} \tag{9.43}$$

where the first term is the sum of the thermal power of the satellites, SAT, and the last term is the thermal power equivalent required to produce the proton beam in the accelerator, ACC. Based on our definitions of symbols, the fission power of a satellite is specifically given by

$$\{P_{\mathrm{fi},t}\}_{\mathrm{SAT},i} = \{U_{\mathrm{Fi}} R_{a,\mathrm{Fi}}\}_{\mathrm{SAT},i} \tag{9.44}$$

while for the accelerator we have

$$\{P_{p,t}\}_{\mathrm{ACC}} = \frac{P_p}{\eta_a} \tag{9.45}$$

where P_p is the proton beam power and η_a is the combined efficiency of converting thermal power into electrical energy and then into the proton beam. Using the results of our analysis of Chapter 4, we write this equation further as

$$\{P_{p,t}\}_{\mathrm{ACC}} = \frac{1}{\eta_a}\left(\frac{I_p E_p}{q_p}\right) \tag{9.46}$$

where I_p is the proton beam current, E_p is the proton energy, and q_p is the proton charge. The thermal system power may hence be written in terms of the more basic systems variables as

$$P_{o,t} = \sum_i \{U_{Fi}R_{a,Fi}\}_{SAT,i} - \frac{I_p E_p}{\eta_a q_p} \tag{9.47}$$

While this describes the interplay of certain variables as they contribute to the system thermal power, it is evidently an insufficient description. If the accelerator breeds fuel only for the satellites then the two terms of Eq. (9.47) are not independent: they are actually coupled by the fissile fuel balance equation. We write therefore for the fissile inventory for the case of fuel cycle equilibrium

$$\frac{d}{dt}N_{Fi}(t) = \sum_i \{R_{c,Fb} - R_{a,Fi}\}_{SAT,i} + \{R_{c,Fb}\}_{ACC} + R^*_{Fi} \tag{9.48}$$

where R^*_{Fi} accounts for the remaining residual reactions involving fissile nuclei. Then, operating at steady state and using the obvious substitution, we obtain

$$\Sigma\{(C_{Fi} - 1)R_{a,Fi}\}_{SAT,i} + \left\{\frac{C_p I_p}{q_p}\right\}_{ACC} \geq 0 \tag{9.49}$$

where C_{Fi} corresponds to the conversion ratio of each satellite. This equation can be used in several ways. For example, for the case of $R^*_{Fi} = 0$ and solving this equation for I_p/q_p, we substitute in Eq. (9.47) and obtain the compact expression

$$P_{o,t} = \sum_i \{P_{fi,t}\}_{SAT,i}\left[1 - \frac{E_p}{U_{Fi}}\frac{(1 - C_{Fi})}{(C_p\eta_a)}\right] \tag{9.50}$$

This equation possesses the form familiar from our fusion–fission systems analysis of previous chapters and can be cast in the form

$$P_{o,t} = \sum_i \{P_{fi,t}\}_{SAT,i}[1 - 1/M_t] \tag{9.51}$$

where a thermal power multiplication parameter has been defined by

$$M_t = \left(\frac{U_{Fi}}{E_p}\right)\frac{C_p\eta_a}{(1 - C_{Fi})} \tag{9.52}$$

In a sense, the spallation accelerator power requirement may evidentl be viewed as analogous to the circulating power used in previous fusion-fission analysis.

9.6. THE SYNERGETIC IDEAL

The preceding discussion lends itself well to some generalization concerning emerging nuclear energy systems options. A particularl appealing option is the selected integration of the fusion and fissio satellites. We indicate in Figure 9.5 how a D–D fusion reactor equippe with a fissile fuel breeding blanket could be integrated with both smal fusion satellites and small fission satellites; the fusion minireactors migh provide electricity by direct conversion of the charged particle reactio products while the fission minireactors generate energy for space-heatin purposes.

From another perspective, we suggest a comparison between conventional isotope flows and those now appearing in a synergetic contex (Figure 9.5). As implied repeatedly throughout this text, this perspectiv must be viewed as a framework rather than a detailed description.

The realization of these expanded isotope flow options would hav substantial implications on system design and operations. As indicated

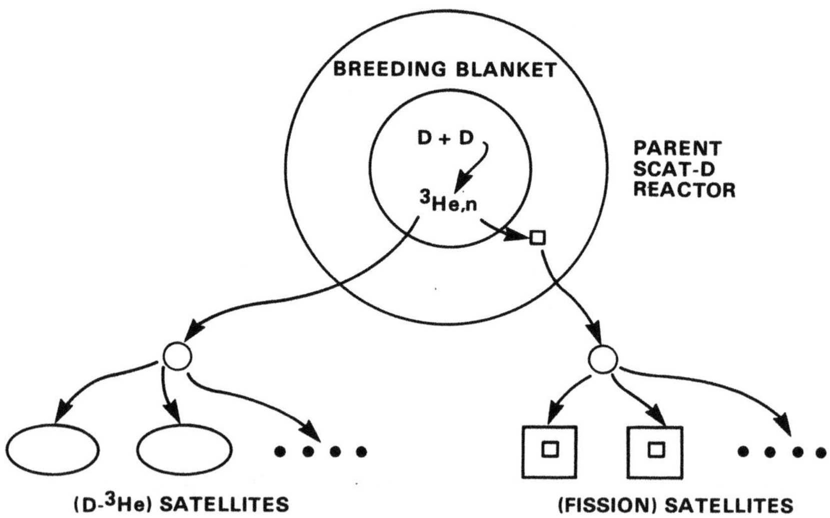

FIGURE 9.5. Illustration suggesting how a D–D parent reactor could supply fu for small fusion and small fission reactors.

ne blanket surrounding a source of neutrons serves three basic and vital unctions:

1. fuel breeding (fissile, fusile);
2. nuclear energy production;
3. selective nuclei destruction.

This itemization encompasses the entire fuel cycle from supply and urning of fuel to management of the reaction products.

ROBLEMS

.1. Investigate the various fusion reactions of Section 9.2 in search of simple and complex fusion chain cycles.

.2. Determine the lower bound for an "acceptable" $\{M_e\}_{PAR}$.

.3. Identify some "merit" parameters which could potentially be used to assess various parent–satellite systems.

.4. Evaluate for the various D–D fusion modes (PURE-D, SCAT-D, CAT-D, EXT-D) the principal figures of merit characterizing their relative usefulness in a synergetic fusion–fission context. Compare D–D-based fusion–fission systems to DT-based ones.

.5. Identify to what extent advanced fusion fuels can be incorporated into a fusion–fission system.

.6. In previous sections mass, energy, and neutron balance equations have been formulated to describe the system characteristics. Extend these formulations to incorporate the destruction of selected waste isotopes (ω) by defining dN_ω/dt in analogy to dN_{Fi}/dt. Specialize for the steady-state case in which $dN/dt = 0$. Extend the definition of $P_{o,e}$ to incorporate the energy requirement for waste destruction.

Chapter Ten

THE EMERGING PERSPECTIVE

In the preceding chapters, we have described and analyzed selected synergetic nuclear energy systems which illustrate the potential utility of the integration of distinct nuclear processes. Our interest in this closing chapter is to examine some implications emerging from this synergetic approach and to identify a broader perspective.

10.1. TRADITION AND INNOVATION

The subject of this text represents one facet of the continuing quest for energy. Such endeavours constitute a history of challenge and ingenuity which are invariably tempered by resource limitations and commercial determination as well as by a widely based societal will. The one identifiable common thread in this enduring enterprise is the necessity of continuing technical adaptation in response to perceived pressures from diverse sources.

The imperative for adaptive change seems to be an invariant of all broadly based technologies. Device improvements and procedural revision are common examples of such change. The less common though more profound adaptations, however, involve a major shift or reorientation of the basic and underlying concepts and processes; quantum jumps in system performance and public acceptability may follow the introduction of such innovations.

It is in such a broader and more interpretive context that we view the civilian nuclear energy program as a technology which seems to be at a threshold for a major reorientation of its dominant emphasis. Indeed, not only the state of technology but also the mood of the times seems opportune to consider a reasonable and feasible extension of the current emphasis on nuclear energy technology in a direction which seems more in agreement with changing societal perceptions and emerging scientific–technical capabilities. We have labeled this emerging perspective *nuclear energy synergism* since it seems to describe well the central feature of integrating various nuclear processes so as to yield advantages not possible otherwise.

Nuclear energy synergism may be characterized as an anticipatory perception of a spectrum of technically feasible systems choices. It is our contention that a variety of choices, not only of specific systems but also of enhanced strategies of development, are in principle possible. In the course of organizing our notions on the subject we recognize the fundamental utility of the accumulated knowledge and experience in nuclear science and technology. Indeed, if there is a distinction with the more traditional view of nuclear energy development, then it rests on the need to adopt a more integrative perspective in systems planning directed more to a synthesis of new systems characteristics from known nuclear processes.

We acknowledge that any emerging framework for a "second generation of nuclear engineering" needs to be developed in considerably more detail before any large-scale commitment to a particular selection could be justified. However, our investigations combined with investigations carried out elsewhere suggest that research on synergetic systems concepts can be made to mesh well with the nuclear power generation and developmental programs carried out to date.

We obviously have a deep interest in the diversity of synergetic nuclear energy options; however, it is not our intent here to express or promote any particular system options or developmental strategy. Those systems which are particularly evident in the scientific–technical marketplace speak for themselves while those which are only describable in a conceptual framework require considerable further analysis. And here we submit that further analysis must involve not only technical issues but also the identification of the role of such a systems choices in a broader temporal and national context.

10.2. NUCLEONICS AND SYSTEMATICS

Implicit in much of our discussion here is the identification of a number of systems characterizations which, in various forms, govern all energy sources but take on particular significance in the present nuclear energy context. Among the various issues requiring consideration we judge the following to be paramount:

(i) Is the reaction density sufficiently high for reasons of economy and functionality?
(ii) Is the recirculation energy component sufficiently small to keep energy losses within satisfactory bounds?
(iii) Is the nuclear fuel available in sufficient quantities?
(iv) Is the program of managing the fuel ashes adequate?

As a minimum, energetics considerations impose the imperative that the total energy necessary to sustain the reaction must be less than the net energy eventually available for societal purposes. The relative readiness of some nuclei to fission upon neutron capture ensures that for such cases the energetic characteristics are indeed most favorable and presently practical. Fusion processes, on the other hand, are evidently more demanding and require continued and substantial additional research effort.

While energetic considerations provide an obvious bound for the useful implementation of a nuclear energy reaction process, no similar bound is generally imposed on the availability of the nuclear fuel. Thus, radiological dangers of reaction products or the strategic importance of nuclear fuels are normally regarded as debatable questions in the implementation of a particular set of nuclear reactions as a source of energy. Such front-end and back-end concerns of the fuel cycle are generally assumed to be manageable by institutional control and regulation. However, it is exactly because of this that the present generation of nuclear engineering suffers in the realm of public acceptability. This problematic of the fuel cycle may—we envisage—be effectively addressed by a synergetic application of selective nuclear processes.

Our nuclear synergism perspective therefore focused in particular on the incorporation of both the front end and the back end of the nuclear fuel process into the entire nuclear energy producing cycle. This required incorporation of additional nuclear processes selected specifically for their effectiveness either in contributing to the intrinsic supply of nuclear fuel or in effecting a reduction of the radioactive nuclear fuel waste. We therefore depict in Figure 10.1 a more inclusive system characterization which represents the basis for an ideal synergetic nuclear energy systems concept. The assessment of alternative nuclear energy systems choices should include consideration of the extent to which it approaches such an ideal.

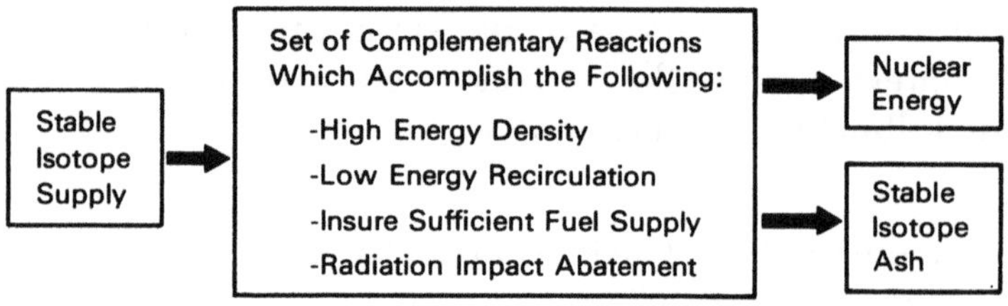

FIGURE 10.1. Listing of essential components of an ideal synergetic nuclear energy system concept.

10.3. NEUTRON CATALYSIS AND MASS–ENERGY SUSTAINABILITY

In our discussion we have referred to $D + T \rightarrow n + \alpha$ as one fusion reaction which may mesh effectively with fission energy technology. While research on D–T fusion represents the most dominant field of present fusion research, subsequent fusion fuels will likely be based on deuterium only. But deuterium-based fusion reactors also emit neutrons and, therefore, like D–T fusion, possess the potential to mesh with fission fuels and fission fuel cycles. This pervasive catalytic role of unbound neutrons within the context of beam/accelerator, fusion, and fission technology thus forms the basis for the enhancement of energy sustainability and fuel mass sustainability. An additional characterization to illustrate nuclear synergism is possible and can be described as follows.

Fission reactors are known to possess a very favorable energy sustainability; by this we mean that only a small fraction of the station lifetime energy output is required to compensate for station operation and initial plant investment. However, except for the fission breeder, thermal fission converter reactors are not isotopically sustainable in the long term because of inadequate natural fissile fuel supply. We plot, therefore, this fission reactor on a graph of energy sustainability vs mass sustainability as shown in Figure 10.2 with detailed definition of the coordinate axes E_s and M_s given in Table 10.1.

Existing beam and accelerator facilities, on the other hand, are essentially unlimited with respect to mass sustainability but are net energy consumers. These technologies can be placed on the lower right part in Figure 10.2.

Present fusion devices require more energy than they produce and also require a supply of tritium which does not exist naturally in sufficient quantities. Hence, the existing state of fusion technology could be placed somewhere in the lower left of Figure 10.2 with the expectation being

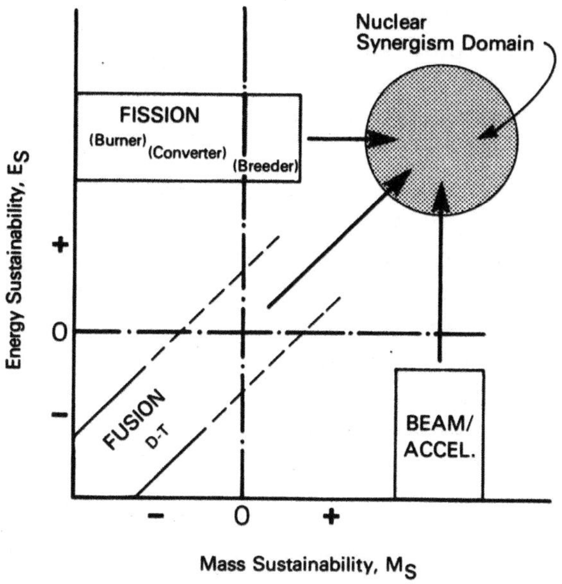

FIGURE 10.2. The placement of existing nuclear technologies on an energy sustainability vs mass sustainability graph is suggested in the text. The circle identifies the nuclear synergism domains and the straight arrows suggest the effective transformation of these technologies associated with a synergistic framework.

that it will soon evolve into the positive isotope and positive energy sustainability domain.

In view of our preceding discussion, the nuclear synergetic domain attainable via neutron catalysis consists of the shaded circle in the upper right part of Figure 10.2; here then the integrated use of reaction and

TABLE 10.1
System Lifetime Definitions of E_s and M_s[a]

Energy sustainability:	
$E_s < 0$:	the system requires more energy than it generates
$E_s = 0$:	the system requires exactly as much energy as it generates
$E_s > 0$:	the system generates more energy than it requires
Mass sustainability:	
$M_s < 0$:	the system destroys nuclear fuel isotopes at an unacceptably high rate
$M_s = 0$:	the system destroys nuclear fuel isotopes at a rate exactly equivalent to its breeding rate
$M_s > 0$:	the system breeds nuclear fuel isotopes at a greater rate than it destroys them or is based on abundant resources

[a] For the case where temporal variations are important, we imply their long term averages.

energy couplings provides, in principle, for both isotope and energy sustainability. Thus, the fusion–fission hybrid and symbiont systems, the spallation-aided fission economy as well as inertial confinement fusion with breeding blankets and similar schemes are contained in this shaded area.

In Figure 10.2 we also illustrate one technology which already operates in the positive isotope and energy sustainability domain. This is the fast fission breeder. Our interpretation of the various synergetic schemes therefore is that they are a form of "breeders" but in a different systems context. While the blanket of a fission reactor or the blanket surrounding a spallation target possesses many of the neutronic properties of a fast breeder, they can also perform additional functions because of the more plentiful supply of neutrons.

10.4. THE NUCLEAR ENERGY CONTINUUM

We indicated in Chapter 1 that the historical nuclear energy record implicitly suggests many possible and distinct options of hybrid and symbiotic nuclear energy systems. A closer examination, however, suggests the basis for an energy system continuum.

Consider first the fusion–fission hybrid. Though it contains both fusion and fission processes, no restrictions on their relative magnitudes are imposed. Indeed, by suitable design and operation, the blanket could serve principally an energy or fuel producing function; that is, it could be an "energy factory" or "fuel factory" or some combination of both. Additionally, if the blanket possesses a low energy gain then the fusion reactions, for a given station power, would dominate and the system could be viewed as a perturbed fusion plant. On the other hand, for a high energy-gain blanket, fissionlike processes are most pronounced and the system appears more like a fission reactor with a novel neutron-production enhancement device. Our point in elaborating on this is that a fusion–fission reactor could in principle be placed anywhere on the rim between the fission and fusion technology spokes of Figure 1.5 depending only upon the relative dominance of the fissionlike or fusionlike processes.

A similar continuum between the beam/accelerator and fission technology can be formulated, with the only difference involving the mechanism for the production of the nonfission neutrons. The continuum characterization between beam/acceleration technology and fusion technology could similarly be represented by the relative integration of these two technologies.

To introduce sufficient generality, we revise Figure 1.5 into a form represented by Figure 10.3, where we also list frequently used names and generic labels associated with current systems concepts.

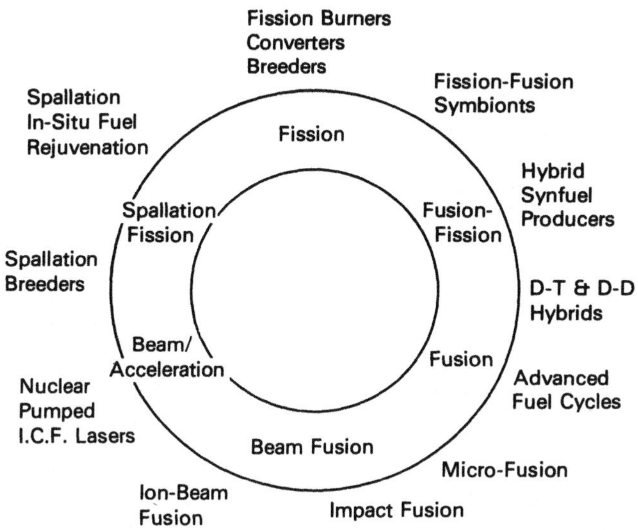

FIGURE 10.3. Expanded depiction of the emerging nuclear energy technology continuum. The listed systems are a partial listing only and meant to illustrate the diversity of existing concepts.

At this juncture it is useful to recognize another intrinsic level of continuity of synergetic nuclear energy systems framework. While the above discussion identified systems which exist specifically between the three reference technologies of Figure 1.5, some emerging nuclear perspectives are best described by spanning a larger part of the rim. For example, both spallation and fusion processes could be viewed as—in a sense—nonfission neutron-producing devices which may contribute to the maintenance of a steady-state neutron–nucleus reaction process in a subcritical fission assembly. This defines, in effect, the domain of driven fission reactors which in terms of our continuum on Figure 10.3 may therefore be placed between the spallation and fusion technology with fission technology at its center, Figure 10.4.

Indeed, we may identify additional more broadly based synergetic concepts. One particularly evident case involves inertial confinement fusion for which the target pellet chamber is surrounded by a fissile/fertile/fusile blanket. That is, the blanket maintains tritium breeding as well as fission and fissile fuel breeding. Clearly, this spans the domain from beam/accelerator technology to fission technology with fusion at the center, Figure 10.4.

Finally, for reason of completeness, we cite nuclear pumped lasers in which fission fragments are used to pump the laser and subsequently induce fusion by inertial confinement. Evidently, this now involves fission and fusion technology with the beam/accelerator technology at its center, Figure 10.4.

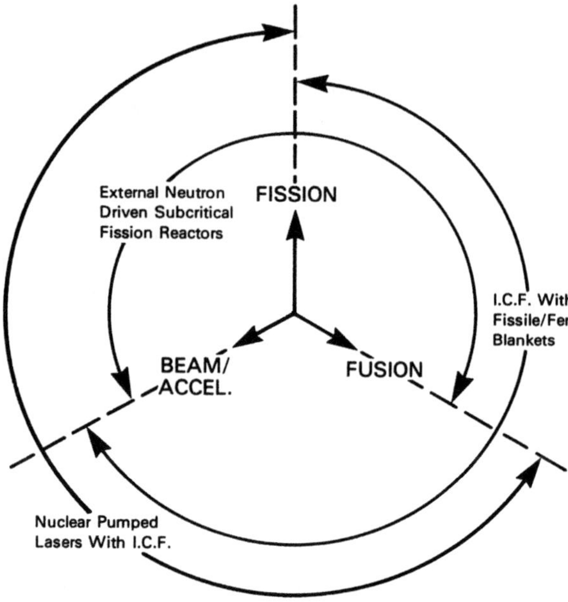

FIGURE 10.4. Suggested representation of breadth of technologies encompassed by particular systems concepts. The abbreviation ICF stands for inertial confinement fusion and could imply the use of lasers or ion beams.

This last example serves the purpose of illustrating another feature in our objective to characterize and demonstrate the existence of a nuclear energy continuum. If, for the case of the nuclear pumped laser, the blanket surrounding the pellet does contain fissile material, then the laser could be placed in the blanket itself—indeed, a special blanket–laser configuration should be conceived—and the fission products from fusion neutron-induced fission would pump the laser. In this case the full cycle of technologies is involved: fusion neutrons lead to fission in the blanket so that the fission products pump the laser so that the laser beam causes fusion by inertial confinement so that the fusion neutrons lead to fission in the blanket . . . indeed, the entire energy system continuum closes on itself.

10.5. AN EXPANDED PERSPECTIVE

Figures 10.2–10.4 constitute a low-dimensional representation of what is essentially a complex multidimensional structure. Thus, in addition to the extent of inclusion of any of the three reference technologies in a synergetic system, assessments specifically associated with subtechnologies of these reference technologies and the overall system function or functions need to be made.

The fission part of a synergetic system may possess a resemblance to existing reactor concepts; the fuel cycles may involve the range from low to high enrichment including the uranium cycle, thorium cycle, or some selected combination. For the fusion part, the systems choices may involve such broad technologies as magnetic or inertial confinement as well as the less emphasized subtechnologies such as colliding beams and impact fusion. The specific system choices may be equally diverse (tokamaks, mirrors, and less conventional types) with fuels initially being D–T and subsequently D–D with its variants, and eventually may also include proton based fuel cycle devices characterized by a reduced neutron yield and expanded direct energy conversion.

The dominant choice of the beam/accelerator part is the driver. Though the two classes are either particles or electromagnetic wave drivers, further choices are extensive. The particles could be electrons, protons, light or heavy ions while the devices could be linacs, cyclotrons, etc. Even for lasers the selection involves wavelength considerations, power capacity, and specific laser types.

The resultant synergetic nuclear energy system may have one or several functions. Among these we cite

1. electricity production,
2. district heating,
3. synthetic fuel production,
4. spent fuel rejuvenation,
5. fission product and actinide destruction, and
6. industrial steam supply.

If the point of view is accepted that fission technology is essentially developed, then a prudent approach would be to render this technology more effective and more useful by developing either spallation accelerators or D–T fusion reactors in an integrative support capacity. Herein, the synergetic nuclear energy ideal could come to a more immediate level of fruition: the fission reactor, being energy rich but neutron poor, would supply some or all of the energy to an accelerator or fusion device and in return receive neutrons to support its energy production and related functions.

While the energy-for-neutron exchange is central, a most useful complimentarity associated with this symbiosis may be identified. This has to do with the feature that since both the spallation and fusion neutrons possess higher neutron source energies than fission neutrons, there exists the physical basis to perform useful functions not possible in a fission reactor domain with its lower neutron source spectrum. The question of interest, therefore, is to determine how these functions can best be realized in a practical configuration.

Basic to the incorporation of a most beneficial fusion–fission–spalla tion symbiosis is the fullest possible exploitation of the blanket whicl surrounds a fusion core or spallation target. Both experimental anc analytical considerations suggest that there are indeed many distinc blanket design options serving a variety of purposes.

10.6. SYNERGISM REVISITED

In the preceding we have attempted to provide a historical, scientific and analytical outline of selected aspects of synergetic nuclear energ systems concepts. It seems fitting to reflect upon some of the fundamenta features of these concepts which identify broader classes of characteri zations.

At one level, nuclear synergism may be viewed as a systemati continuation of the intellectual tradition which seeks to synthesize an adapt technology to changing conditions. The fundamental feature of thi emphasis then deserving of wider recognition is its relation to that aspec of human stewardship which seeks to establish an enhanced matchin between the needs of a community and the imperative of its broade environment.

Reflecting on the nuclear implications, synergism relates specificall to the means of increasing the number of neutrons in order to attai enhanced nuclear fuel management prospects for nuclear energy release In such a context, nuclear synergism is then a variant of the early pioneer ing ideas from the civilian nuclear power program. It will be recalle that it was the growing realization of the scarcity of naturally occurrin fissile nuclei which served as the motivation to develop fast fissio breeders and thermal near-breeders as well as fusion reactors. Viewed i this perspective, nuclear synergism is a descendant of these early idea not only in thought but also in substance. The consequence of this notio of continuity is that the development of synergetic systems does nc require a redirection of basic purpose but only a redirection c approach.

At another level, we point to the many examples of complimentarit existing in natural and mature man-made systems. What may be a by product or waste output in one part may serve as a necessary inpu ingredient to another. Where one part of a system is deficient in on function but excels in another, one can enhance the overall performanc by a suitable integration with another part whose efficiency in the sam two functions is reversed. Such a complimentary duality finds its syne getic nuclear equivalence in the integration of "neutron rich but energ poor" processes with those which are "neutron poor but energy rich.

An inherent feature of nuclear synergetics is therefore the provision for energy and isotope exchange linkages.

As a final point of reflection, we return to the initial triad of nuclear technologies and consider the fundamental interplay between mass and energy as it relates to current technology and developmental expectations. To aid in this perspective, we refer again to Figure 10.2. It is evident that even in the absence of any synergetic considerations, the present dominant focus of research and development in fusion, fission, and beam/accelerator technology is in the direction towards enhanced energy and mass sustainability. That is, the driving motivation is to transform the technology into the positive quadrant of the E_s–M_s graph. The overall viability of the system must, in some measure, be governed by the degree of penetration into this positive $E_s > 0$ and $M_s > 0$ domain.

It is remarkable then that the general aims of energy and mass sustainability of stand-alone fission or fusion energy systems can, apparently, with greater ease be reached by a suitable synthesis of the three nuclear technologies; fission, fusion, and beam/accelerators.

PROBLEMS

10.1. Investigate selected nuclear energy systems concepts listed in Figure 10.3 which are not generally known (e.g., nuclear pumped lasers, impact fusion, etc.).

10.2. Superimposed on the energy continuum, one may also impose the parameter of "neutron richness" or "neutron poverty" on a per reaction basis. Plot this characterization taking care to be specific about the selected reaction.

10.3. It would be desirable to add another Cartesian coordinate to Figure 10.2 possibly labeled *system resilience*. Attempt a characterization of various systems based on this third axis.

Appendix A
Bibliography

This bibliography has been assembled as an aid to the further study of synergetic nuclear energy systems and related topics. With the exception of several early sources, the references listed here are from the refereed or generally available journal literature only and are ordered by year and alphabetically by lead author within each year. The period covered is from 1952 through 1981.

1. W. B. Lewis, *The Significance of the Yield of Neutrons from Heavy Elements Excited to High Energies*, Chalk River Nuclear Laboratories, Chalk River, Canada, DR-24 (1952).
2. F. Powell, *Proposal for a Driven Thermonuclear Reactor*, California Research and Development Company, United States Atomic Energy Commission, Report No. LWS-24920 (1953).
3. J. D. Lawson, *A Survey of Some Suggested Methods of Realising Fusion Reactors*, Atomic Energy Research Establishment, Harwell, U.K., AERE-GP/M No. 185 (1955).
4. H. Jung, Two suggestions regarding controlled fusion: approximation equations for fusion–fission reactors and fusion-reaction-tube reactors, *Nuclear Fusion*, **9**, 169 (1969).
5. L. M. Lidsky, *Fission–Fusion Symbiosis: General Considerations and a Specific Example*, Proceedings of Nuclear Fusion Reactors Conference, 17–19 September 1969, Culham Laboratory, U.K. (1969).
6. V. A. Davidenko, On electronuclear breeding, *Soviet Atomic Energy*, **29**, 866 (1970).
7. D. Steiner, The nuclear performance of fusion reactor blankets, *Nuclear Application and Technology*, **9**, 83 (1970).
8. V. G. Vasilkov *et al.*, The electronuclear method of generating neutrons and producing fissionable materials, *Soviet Atomic Energy*, **29**, 858 (1970).
9. D. West and E. Ward, Neutron yield from protons and deuterons of momenta between 0.8 and 1.7 GeV/*c* totally absorbed in lead, *Canadian Journal of Physics*, **15**, 2061 (1971).
10. R. E. Kay *et al.*, SLOWPOKE: a new low-cost laboratory reactor, *International Journal of Applied Radiation and Isotopes*, **24**, 509 (1973).

11. B. R. Leonard, Jr., A review of fusion–fission (hybrid) concepts, *Nuclear Technology*, **20**, 161 (1973).
12. C. Powell and D. J. Hahn, Energy balance of a hybrid fusion–fission reactor, *Atomkernenergie*, **21**, 172 (1973).
13. V. S. Barashenkov, F. D. Tonee, and S. E. Chigrinov, Interaction of high-energy deuteron beam in matter, *Soviet Atomic Energy*, **37**, 1256 (1974).
14. R. N. Horoshko, H. Hurwitz, and H. Zmora, Application of laser fusion to the production of fissile materials, *Annals of Nuclear Science and Engineering*, **1**, 223 (1974).
15. Y. I. Kolesnichenko and S. N. Reznik, D–T plasma as a source of neutrons for the combustion of uranium-238, *Nuclear Fusion*, **14**, 114 (1974).
16. P. Fortescue, The fusion breeder concept, *Annals of Nuclear Energy*, **2**, 29 (1975).
17. I. N. Golovin, The position of hybrid reactors in power-generation systems, *Soviet Atomic Energy*, **39**, 1035 (1975).
18. C. W. Gordon and A. A. Harms, Comparative energetics of three fusion–fission symbiont nuclear reactor systems, *Nuclear Engineering and Design*, **34**, 296 (1975).
19. A. A. Harms, Hierarchical systematics of fusion–fission energy systems, *Nuclear Fusion*, **15**, 939 (1975).
20. G. LaVerque, J. E. Robinson, and J. G. Martel, On the matching of fusion breeders to heavy-water reactors, *Nuclear Technology*, **26**, 12 (1975).
21. L. M. Lidsky, Fusion–fission systems: hybrid, symbiotic and augean, *Nuclear Fusion*, **15**, 151 (1975).
22. N. G. Cook and J. A. Maniscalco, Uranium-233 breeding and neutron multiplying blankets for fusion reactors, *Nuclear Technology*, **30**, 5 (1976).
23. A. A. Harms, Upper bounds of fissile fuel yield with fusion breeders, *Canadian Journal of Physics*, **54**, 1637 (1976).
24. A. A. Harms and C. W. Gordon, Fissile fuel breeding potential with paired fusion–fission reactors, *Annals of Nuclear Energy*, **3**, 411 (1976).
25. Y. I. Kolesnichenko and S. N. Resnik, The D–D nuclear fusion reaction in a hybrid reactor, *Nuclear Fusion*, **16**, 97 (1976).
26. J. Maniscalco, Fusion–fission hybrid concepts for laser induced fusion, *Nuclear Technology*, **28**, 98 (1976).
27. W. Seifritz, The symbiosis between beam-driven hybrid DT-fusion reactors and near-breeder HTGR's, *Atomwirtschaft*, **21**, 205 (1976).
28. S. Shang-Fou, G. L. Woodruff, and N. J. McCormick, A high-gain fusion–fission reactor producing uranium-233, *Nuclear Technology*, **29**, 392 (1976).
29. J. Davies, Conserving uranium without the fast breeder, *Nature*, **270**, 376 (1977).
30. C. W. Gordon and A. A. Harms, The basic characteristics of an efficient fusion breeder, *Atomkernenergie*, **29**, 213 (1977).
31. A. A. Harms and C. W. Gordon, A parametric analysis of the spallation breeder, *Nuclear Science and Engineering*, **63**, 336 (1977).
32. B. C. Maglich, Ed., *Proc. First Symposium on Clean Fusion, Nuclear Instruments and Methods*, **144** (1), (1977).
33. H. Nakashima, M. Ohta, and T. Seki, Nuclear characteristics of D–D fusion reactor blankets (I), *Journal of Nuclear Science and Technology*, **14**, 75 (1977).
34. H. A. Bethe, The fusion hybrid, *Nuclear News*, **27**, 41 (May 1978).
35. V. L. Blinkin and V. M. Novikov, Symbiotic system of a fusion and fission reactor with very simple fuel reprocessing, *Nuclear Fusion*, **18**, 7 (1978).
36. L. P. Feoktistov *et al.*, Hybrid reactor based on laser-induced thermonuclear fusion *Soviet Journal of Quantum Electronics*, **8**, 201 (1978).
37. E. Greenspan *et al.*, Source driven breeding thermal power reactors using D–T fusion neutron sources, *Atomkernenergie*, **32**, 30 (1978).
38. E. Greenspan, Source driven breeding thermal power reactors using lithium free neutron sources, *Atomkernenergie*, **32**, 133 (1978).
39. A. A. Harms, Nuclear energy synergetics, *Atomkernenergie*, **32**, 3 (1978).
40. A. A. Harms and W. T. Hartmann, Spent nuclear fuel reenrichment without reprocessing, *Annals of Nuclear Energy*, **5**, 213 (1978).

41. A. A. Harms and M. Heindler, The matching of dense plasma focus devices with fission reactors, *Nuclear Science and Engineering*, **66**, 1 (1978).
42. M. Heindler, H. Lang, and A. A. Harms, The role of nonfission neutron sources in a fission reactor economy, *Atomkernenergie*, **32**, 142 (1978).
43. J. D. Lee, Mirror fusion–fission hybrids, *Atomkernenergie*, **32**, 19 (1978).
44. J. R. McNally Jr., D–^{3}He as a clean fusion reactor, *Nuclear fusion*, **18**, 133 (1978).
45. G. H. Miley, Advanced fuel fusion concepts, *Atomkernenergie*, **32**, 12 (1978).
46. H. Nahashima, M. Ohta, and Y. Nakao, Nuclear characteristics of D–D fusion reaction blankets (II), *Journal of Nuclear Science and Technology*, **15**, 490 (1978).
47. S. O. Schriber, Electronuclear fuel production using high-intensity accelerators, *Atomkernenergie*, **32**, 49 (1978).
48. G. Shani, Tailoring the breeding–fission ratio in a hybrid fusion reactor blanket by neutron moderation with iron, *Nuclear Science and Engineering*, **65**, 183 (1978).
49. M. Steinberg *et al.*, Electronuclear fissile fuel production, *Atomkernenergie*, **32**, 39 (1978).
50. G. Velarde *et al.*, Analysis of laser–fission–fusion systems: time-dependent coupled nuclear-thermohydrodynamic analysis and application, *Atomkernenergie*, **32**, 58 (1978).
51. E. P. Velikhov *et al.*, Tokamak-type thermonuclear hybrid reactor for the production of fission fuel and electric energy, *Soviet Atomic Energy*, **45**, 653 (1978).
52. D. J. Bender, Performance parameters for fusion–fission power systems, *Nuclear Technology*, **44**, 381 (1979).
53. P. Grant, The use of high energy accelerators in the nuclear fuel cycle, *Nature*, **278**, 693 (1979).
54. M. R. Jonzen, Mass–energy analysis for gas-cooled fast reactors and fusion–fission hybrid reactor systems, *Nuclear Technology*, **45**, 54 (1979).
55. A. S. Kukushkin and V. I. Pistumovich, Optimization of the plasma parameters in a hybrid-reactor–Tokamak, *Soviet Atomic Energy*, **47**, 983 (1979).
56. G. H. Miley and J. G. Gilligan, A possible route to small, flexible fusion units, *Energy*, **4**, 163 (1979).
57. K. F. Schoepf and A. A. Harms, Characteristics of a beam-driven deuterium fuelled fusion–fission reactor system, *Nuclear Science and Engineering*, **71**, 170 (1979).
58. K. F. Schoepf and A. A. Harms, The synergetics of the catalytic D–D fusion breeder, *Nuclear Fusion*, **19**, 5 (1979).
59. V. G. Vasil'kov *et al.*, The electronuclear method of generating neutrons and producing fissionable materials, *Soviet Atomic Energy*, **29**, 858 (1979).
60. S. I. Abdel-Kalik *et al.*, Impact of fusion–fission hybrids on world nuclear futures, *Atomkernenergie/Kerntechnik*, **36**, 23 (1980).
61. V. C. Badham *et al.*, Accidental criticality of a fusion–fission hybrid blanket design, *Nuclear Technology*, **47**, 221 (1980).
62. Y. G. Bobkov *et al.*, Theoretical studies on the accumulation of ^{232}U, ^{236}Pu, and ^{238}Pu in the breeding zones of hybrid and fast reactors, *Soviet Atomic Energy*, **48**, 407 (1980).
63. R. W. Conn *et al.*, Hybrids for direct enrichment and self-protected fissile fuel production, *Nuclear Technology*, **49**, 458 (1980).
64. A. A. Harms, Synergetic nuclear energy systems concepts, *Advances in Nuclear Science and Technology*, **12**, 211 (1980).
65. A. A. Harms and J. Ligou, The systematics of emerging nuclear energy concepts, *Atomkernenergie/Kerntechnik*, **36**, 3 (1980).
66. M. Heindler and A. A. Harms, Efficiency merit assessment of emerging synergetic energy systems, *Atomkernenergie/Kerntechnik*, **36**, 7 (1980).
67. J. Jung, A comparative study of tritium breeding performance of lithium, Li_2O, and Li_7Pb_2 blankets in a Tokamak power reactor, *Nuclear Technology*, **50**, 60 (1980). (1980).
68. J. D. Lee, Tandem mirror fusion–fission hybrid studies, *Atomkernenergie/Kerntechnik*, **36**, 36 (1980).
69. H. Makowitz *et al.*, Hyperfuse: a novel inertial confinement system utilizing hypervelocity projectiles for fusion energy production and fission waste transmutation, *Atomkernenergie/Kerntechnik*, **36**, 47 (1980).

70. V. M. Novikov *et al.*, Experimental study of the Pb and Bi breeding zones effect on the neutron parameters of a molten salt blanket in a fusion reactor, *Nuclear Instruments and Methods*, **173**, 449 (1980).
71. J. K. Presley, A. A. Harms, and M. Heindler, Nuclear fuel trajectories of fusion–fission symbionts, *Nuclear Science and Engineering*, **74**, 193 (1980).
72. M. M. H. Ragheb *et al.*, Nuclear performance of molten salt fusion–fission symbiont systems for catalyzed deuterium–deuterium and deuterium–tritium reactors, *Nuclear Technology*, **48**, 216 (1980).
73. S. Sahin *et al.*, Basic structure of the fusion–fission (hybrid) reactor, *Atomkernenergie/Kerntechnik*, **36**, 33 (1980).
74. K. F. Schoepf, Beam driven D-fusion plasma with a fusion–fission hybrid system, *Atomkernenergie/Kerntechnik*, **36**, 36 (1980).
75. D. V. Markovskii and G. E. Shatalov, Sensitivity of characteristics of hybrid reactors to spectra of secondary neutrons, *Soviet Atomic Energy*, **42**, 509 (1980).
76. M. Steinberg *et al.*, The linear accelerator fuel enricher regenerator (LAFER) and fission product transmuter (APEX), *Atomkernenergie/Kerntechnik*, **36**, 42 (1980).
77. A. S. Tai and R. A. Krakowski, An economic parametric analysis of the synthetic fuel production by a fusion–fission complex, *Atomkernenergie/Kerntechnik*, **36**, 18 (1980).
78. G. L. Woodruff, Neutronics performance of a flowing particle bed for synthetic fuel production in fusion–fission hybrid reactors, *Atomkernenergie/Kerntechnik*, **36**, 13 (1980).
79. M. Z. Youssef *et al.*, Tritium and fissile fuel exchange between hybrids, fission power reactors, and tritium production reactors, *Nuclear Technology*, **47**, 397 (1980).
80. D. H. Berwald and J. A. Maniscalco, An economic method for symbiotic fusion–fission electricity generation systems, *Nuclear Technology/Fusion*, **1**, 128 (1981).
81. D. H. Berwald and J. A. Maniscalco, Performance and economics analysis of several laser fusion breeder fueled electricity generation systems, *Nuclear Technology/Fusion*, **1**(1), 137 (1981).
82. R. W. Conn *et al.*, Fusion–fission hybrid design with analysis of direct enrichment and nonproliferation features, *Nuclear Engineering and Design*, **63**(2), 357 (1981).
83. A. A. Harms and W. Haefele, Nuclear Synergism, *American Scientist*, **69**(3), 310 (1981).
84. D. L. Jassby, The fusion-supported decentralized nuclear energy system, *J. Fusion Energy*, **1**, 59 (1981).
85. J. L. Kelly and R. P. Rose, The Tokamak hybrid reactor, *Nuclear Engineering and Design*, **63**, 395 (1981).
86. E. M. Krenciglowa and A. A. Harms, An assessment of nuclear waste transmutation with proton accelerators, *Nuclear Instruments and Methods*, **185**, 393 (1981).
87. R. W. Moir, The tandem mirror hybrid reactor, *Nuclear Engineering and Design*, **63**, 375 (1981).
88. K. R. Schultz, Gas-cooled fusion–fission hybrid reactor system, *J. Fusion Energy*, **1**, 163 (1981).
89. R. P. Rose, The case for the fusion hybrid, *J. Fusion Energy*, **1**, 185 (1981).
90. J. P. Holdren, Fusion–fission hybrids: environmental aspects and their role in hybrid rationale, *J. Fusion Energy*, **1**, 197 (1981).
91. S. I. Abdel-Khalik, P. Jansen, G. Kessler, and P. Klumpp, Impact of fusion–fission hybrids on the world nuclear future, *Atomkernenergie/Kerntechnik*, **38**, 1 (1981).
92. E. Greenspan and G. H. Miley, Fissile and synthetic fuel production ability of hybrid reactors, *Atomkernenergie/Kerntechnik*, **38**, 12 (1981).
93. V. L. Blinkin and V. M. Novikov, On the choice of economical optimum blanket of hybrid reactors, *Atomkernenergie/Kerntechnik*, **38**, 20 (1981).
94. H. Nakashima, M. Ohta, and T. Yasutake, Strontium-90 transmutation in D–D fusion reactors, *Atomkernenergie/Kerntechnik*, **38**(1), 27 (1981).
95. H. Makowitz, J. R. Powell, and R. Wiswall, Hyperfuse: a hypervelocity inertial confinement system for fusion energy production and fission waste transmutation, *Nucl. Sci. Eng.* **78**(4), 395 (1981).
96. S. Taczanowski, Neutron multiplier alternatives for fusion reactor blankets, *Annals Nuclear Energy*, **8**(1), 29 (1981).

97. J. A. Maniscalco, D. H. Berwald, R. B. Campbell, R. W. Moir, and J. D. Lee, Recent progress in fusion–fission hybrid reactor design studies, *Nuclear Technology/Fusion*, 1(4), 419 (1981).
98. K. Erickson, N. J. McCormick, and G. L. Woodruff, Power density flattening in fusion–fission hybrid reactors, *Nuclear Technology/Fusion*, 1(4), 533 (1981).
99. H. Nakashima, S. Kuroki, H. Rikihisa, and M. Ohta, Fuel cycle characteristics of base–satellite fusion reactor systems, *Atomkernenergie/Kerntechnik*, 38(3), 161 (1981).
100. T. A. Parish, On the inventory dynamics of batch-fueled fusion–fission systems, *Nucl. Science and Engineering* 79(3), 299 (1981).
101. A. A. Harms, E. M. Krenciglowa, W. W. Sassin, and J. W. Hilborn, Accelerator-supported regionally distributed nuclear heat systems with nuclear waste abatement capacity, *Annals Nuclear Energy*, 8, 431 (1981).
102. S. Sahin, Neutronic analysis of fast hybrid thermionic reactors, *Atomkernenergie/Kerntechnik*, 39, 41 (1981).
103. M. M. H. Ragheb, Optimized fissile and fusile breeding in a laser–fusion fissile-enrichment flux trap blanket, *J. Fusion Energy*, 1, 285 (1981).
104. J. D. Lee and R. W. Moir, Fusion-suppressed blankets for fissile fuel breeding fusion reactors, *J. Fusion Energy*, 1, 299 (1981).

Appendix B

Nuclear Reaction Characterization

The characterization of nuclear reactions in our context requires consideration of various processes and a very large range of parameters. For example, in the case of fusion, both particles are charged and in motion. In fission, one of the interacting particles is essentially stationary while the other is neutral and may possess kinetic energy in a range spanning up to 8 orders of magnitude. In some cases, such as D–D fusion, some interactions involve identical particles.

The diversity of physical processes and the range of parameters notwithstanding, it is possible to characterize the nuclear reactions of interest in a general form with the aid of the cross-section concept. This approach possesses a geometrical as well as physical basis and is of broad utility. To develop a sufficiently comprehensive reaction characterization involving nucleons and nuclei we consider the following.

The number density of the i-type particles will be represented by n_i and possesses units of particles per m^3. Taking this number density to be uniform in a given reaction volume V yields the total number of particles N_i as

$$N_i = n_i V \tag{B.1a}$$

$$N_i = \int_V M_i(\mathbf{x})\, d\mathbf{x} \tag{B.1b}$$

Here, the particle density n_i will be distinguished from the symbol for a neutron n by the subscript which identifies the type of particle.

Consider now an arbitrary region of volume V containing a uniform particle densities n_A and n_B of A- and B-type particles with $\mathbf{v}_A$ and $\mathbf{v}_B$

their velocity vectors with respect to the laboratory system. The probability of a reaction occurring is proportional to the product of the respective densities of A-type and B-type particles and their relative speed

$$v_{\text{rel}} = |\mathbf{v}_A - \mathbf{v}_B| \tag{B.2}$$

This suggests that the reaction rate density, r_{AB} for short range and some long range reactions, has a form given by a proportionality relationship

$$r_{AB}(v_{\text{rel}}) \propto v_{\text{rel}} n_A n_B \tag{B.3}$$

A zero value for any of these three factors on the right will ensure a zero reaction rate density. A scale change in any one of them, with the remaining factors constant, will lead to a corresponding change in the magnitude of the reaction density. The constant of proportionality required to transform this relationship into an equation is called the microscopic cross section:

$$r_{AB}(v_{\text{rel}}) = \sigma_{AB}(v_{\text{rel}}) v_{\text{rel}} n_A n_B \tag{B.4}$$

A value of 10^{-28} m^2 for $\sigma_{AB}(v_{\text{rel}})$ is the conventional unit for the microscopic nuclear cross sections called a barn. Note the dependence of this cross section on the relative speed.

Nuclear particles can interact in many different ways: fission, fusion, scattering, capture, etc. Each of these processes possesses its own cross section which we will identify by the subscript r:

$$r_{r,AB}(v_{\text{rel}}) = \sigma_{r,AB}(v_{\text{rel}}) v_{\text{rel}} n_A n_B \tag{B.5}$$

For the more general space-dependent case of both A-type and B-type particles having an arbitrary distribution in velocity space described by the macroscopic single particle velocity distribution functions $f_A(\mathbf{v}_A, \mathbf{x})$ and $f_B(\mathbf{v}_B, \mathbf{x})$, we obtain

$$\begin{aligned} r_{r,AB}(\mathbf{x}) &= \int_{\mathbf{v}_A} \int_{\mathbf{v}_B} r_{r,AB}(v_{\text{rel}})\, d\mathbf{v}_B\, d\mathbf{v}_A \\ &= n_A(\mathbf{x}) n_B(\mathbf{x}) \int_{\mathbf{v}_A} \int_{\mathbf{v}_B} \sigma_{r,AB}(v_{\text{rel}}) v_{\text{rel}}\, f_A(\mathbf{v}_A, \mathbf{x}) f_B(\mathbf{v}_B, \mathbf{x})\, d\mathbf{v}_B\, d\mathbf{v}_A \end{aligned} \tag{B.6}$$

as the general expression for the reaction rate density. The differentials $d\mathbf{v}_A \, d\mathbf{v}_B$ define a volume element in the six-dimensional velocity space and the distributions $f_i(\mathbf{v}_i, \mathbf{x})$ are taken to be normalized:

$$\int_{\mathbf{v}} f_i(\mathbf{v}_i, \mathbf{x}) \, d\mathbf{v}_i = 1 \tag{B.7}$$

The relationship Eq. (B.6) is complete and explicit if the particles A and B are distinct and if only interactions involving the A-type particles with B-type particles occur. Special care needs to be exercised if the A-type particles and/or B-type particles interact also with themselves or when there is only one type of self-interacting particle as we illustrate next.

For the $A \neq B$ reaction case, the product $n_A n_B$ represents an accounting of any one of the A particles, say the ith, interacting with any one of the B particles, say the jth. Hence we represent each possible interaction as an element of a matrix given by

$$(n_A, n_B)_{ij} = (i, j), \qquad i = 1, 2, \ldots, n_A, \qquad j = 1, 2, \ldots, n_B \tag{B.8}$$

and the number of possible and distinct interactions is the number of matrix elements. Now if $A = B$, then $n_A = n_B = n_0$, and the diagonal matrix elements (i, i) must be excluded from the enumeration since a particle cannot interact with itself. The remaining $(n_0^2 - n_0)$ pairings include the transposed elements (i, j) and (j, i), each pair describing the same interaction between the ith and jth particles; therefore only the elements on one side of the diagonal need to be considered, yielding a total of $(n_0^2 - n_0)/2$ interactions. Since in all cases of interest, n_i is very large, we may simplify this result to

$$\frac{n_0^2 - n_0}{2} = \frac{1}{2} n_i (n_i - 1) \simeq \frac{n_0^2}{2} \tag{B.9}$$

Eq. (B.6), augmented by Eq. (B.9) as necessary, is the essential form for the reaction rate density which can be fully generalized by the introduction of the Kronecker delta with indices A and B

$$r_{r,AB}(\mathbf{x}) = \frac{n_A(\mathbf{x}) n_B(\mathbf{x})}{1 + \delta_{AB}} \int_{\mathbf{v}_A} \int_{\mathbf{v}_B} \sigma_{r,AB}(v_{\text{rel}}) v_{\text{rel}} f_A(\mathbf{v}_A, \mathbf{x}) f_B(\mathbf{v}_B, \mathbf{x}) \, d\mathbf{v}_B \, d\mathbf{v}_A \tag{B.10}$$

We now consider its specialization to particularly relevant cases of interac[-]
tions.

In fusion physics it is common to take the two velocity distributio[n] functions as essentially space independent. Then, one defines the functio[n]

$$\langle \sigma v \rangle_{AB} = \int_{\mathbf{v}_A} \int_{\mathbf{v}_B} \sigma_{\mathrm{fu},AB}(v_{\mathrm{rel}}) v_{\mathrm{rel}} f_A(\mathbf{v}_A) f_B(\mathbf{v}_B)\, d\mathbf{v}_A\, d\mathbf{v}_B \qquad \text{(B.11)}$$

frequently called the fusion reactivity or fusion reaction parameter wit[h] $\sigma_{\mathrm{fu},AB}(v_{\mathrm{rel}})$ specifically the fusion cross section. Equation (B.10) may henc[e] be written in the compact form

$$r_{\mathrm{fu},AB}(\mathbf{x}) = \frac{n_A(\mathbf{x}) n_B(\mathbf{x})}{1 + \delta_{AB}} \langle \sigma v \rangle_{AB} \qquad \text{(B.12)}$$

The reaction parameter $\langle \sigma v \rangle_{AB}$ is generally calculated for the case tha[t] $f_A(\mathbf{v}_A)$ and $f_B(\mathbf{v}_A)$ are represented by a Maxwellian distribution with th[e] kinetic temperature as a distribution parameter. In the case that [A] and B particles are characterized by the same temperature T, the averag[e] particle energy for both particles is $(3/2)kT$, where k is the Boltzma[nn] constant and T is the absolute temperature; the most probable particl[e] energy is kT. The reactivity parameter $\langle \sigma v \rangle_{AB}$ is generally displayed as [a] function of kT in units of eV.

Different considerations enter into a convenient formulation of Eq[.] (B.6) for neutron–nucleus interactions. Here, we let particle $A = n$ be [a] neutron and particle $B = F$ an atomic nucleus. A satisfactory impositio[n] is that the velocity distribution of the nucleus not be a function of positio[n] while the neutron velocity distribution does depend on the position $\mathbf{x}$ i[n] the volume of interest. Hence, we redefine

$$f_A(\mathbf{v}_A, \mathbf{x}) \rightarrow f_n(\mathbf{v}_n, \mathbf{x}) \qquad \text{(B.13)}$$

and

$$f_B(\mathbf{v}_B, \mathbf{x}) \rightarrow f_F(\mathbf{v}_F) \qquad \text{(B.14)}$$

where the latter characterizes the thermal motion of an atom about its equilibrium lattice point. We may write Eq. (B.6) in the general form

$$r_{r,nF}(\mathbf{x}) = n_F(\mathbf{x}) n_n(\mathbf{x}) \int_{\mathbf{v}_n} f_n(\mathbf{v}_n, \mathbf{x}) \left\{ \int_{\mathbf{v}_F} \sigma_{r,nF}(v_{\mathrm{rel}}) v_{\mathrm{rel}} f_F(\mathbf{v}_F)\, d\mathbf{v}_F \right\} d\mathbf{v}_n \qquad \text{(B.15)}$$

where, for reasons of completeness, a space-dependent neutron and nucleus density has been explicitly noted. Next we define an effective microscopic cross section $\tilde{\sigma}_{r,nF}$ for an r-type interaction between a neutron and a nucleus by the so-called Doppler broadening integral involving the velocity distribution of the F nuclei and the microscopic cross section

$$\tilde{\sigma}_{r,nF}(v_n) = \frac{1}{v_n} \int_{\mathbf{v}_F} \sigma_{r,nF}(v_{\text{rel}}) v_{\text{rel}} f_F(\mathbf{v}_F) \, d\mathbf{v}_F \tag{B.16}$$

Then the neutron–nucleus interaction rate is given by

$$r_{r,nF}(\mathbf{x}) = n_F(\mathbf{x}) n_n(\mathbf{x}) \int_{\mathbf{v}_n} f_n(\mathbf{v}_n, \mathbf{x}) \tilde{\sigma}_{r,nF}(v_n) \nu_n \, d\mathbf{v}_n \tag{B.17}$$

While this equation is complete and correct, a special notation has emerged in fission reactor analyses. The neutron kinetic energy in the laboratory system, E, has become widely used as the independent variable. Hence, to transform from the vector space $\mathbf{v}_n$ we introduce the unit directional vector $\mathbf{\Omega}$ as

$$\mathbf{\Omega} = \frac{\mathbf{v}_n}{|\mathbf{v}_n|} \tag{B.18}$$

and define the neutron vector flux by

$$\Phi(\mathbf{x}, E, \mathbf{\Omega}) \, dE_n \, d\mathbf{\Omega} = n_n(\mathbf{x}) f_n(\mathbf{v}_n, \mathbf{x}) v_n \, d\mathbf{v}_n \tag{B.19}$$

Furthermore, a macroscopic cross section is defined and related to the effective microscopic cross section by

$$\Sigma_{r,nF}(E, \mathbf{x}) = \sigma_{r,nF}(v_n) n_F(\mathbf{x}) \tag{B.20}$$

Inserting therefore Eq. (B.19) and Eq. (B.20) into Eq. (B.17) yields the neutron–nucleus reaction rate density

$$r_{r,nF}(\mathbf{x}) = \int_E \int_{\mathbf{\Omega}} \Sigma_{r,nF}(E, \mathbf{x}) \Phi(E, \mathbf{x}, \mathbf{\Omega}) \, d\mathbf{\Omega} \, dE_n$$

$$= \int_E \Sigma_{r,nF}(E, \mathbf{x}) \Phi(E, \mathbf{x}) \, dE \tag{B.21}$$

where the scalar neutron flux $\Phi(\mathbf{x}, E_n)$ is defined by

$$\Phi(E, \mathbf{x}) = \int_{\mathbf{\Omega}} \Phi(E, \mathbf{x}, \mathbf{\Omega})\, d\mathbf{\Omega} \tag{B.22}$$

The subscript r in Eq. (B.21) stands for processes such as fission, neutron capture, etc.

Spallation reactions involve an essentially stationary high-Z nucleus and an energetic proton and may be described analogously to a neutron–nucleus interaction in a fission reactor. We therefore use the notation $A = p$ for the proton and $B = Z$ for the heavy nucleus. Proceeding as we did for neutron–nucleus interaction and changing the symbolism of Eq. (B.17) yields

$$r_{pZ}(\mathbf{x}) = n_Z(\mathbf{x}) n_p(\mathbf{x}) \int_{\mathbf{v}_p} \tilde{\sigma}_{pZ}(\mathbf{v}_p) f_p(\mathbf{v}_p, \mathbf{x}) v_p\, d\mathbf{v}_p \tag{B.23}$$

where a space dependence on the proton density $n_p(\mathbf{x})$ is indicated.

A number of substitutions could now be introduced. The most obvious is to transform to the kinetic energy of the proton, E_p. This then yields, as one convenient expression

$$r_{pZ}(\mathbf{x}) = n_Z(\mathbf{x}) n_p(\mathbf{x}) \int_{E_p} \tilde{\sigma}_{pZ}(E_p) f_p(E_p, \mathbf{x})\, dE_p \tag{B.24}$$

While the above expressions r_{AB}, $r_{r,nF}$, and r_{pZ} represent reaction rate densities, possessing units of $(\mathrm{m}^3\ \mathrm{s})^{-1}$, the total number of interactions per unit time in a specified volume V is given by

$$R_{r,AB} = \int_V r_{r,AB}(\mathbf{x})\, d\mathbf{x} \tag{B.25}$$

In summary, then, the three pertinent nuclear reactions of main interest here, expressed as reaction rate densities or total reaction rates in a form generally used, are the following:

(a) Fusion reactions involving particles A and B:

$$r_{AB}(\mathbf{x}) = \frac{n_A(\mathbf{x}) n_B(\mathbf{x})}{1 + \delta_{AB}} \langle \sigma v \rangle_{AB} \tag{B.26a}$$

$$R_{AB} = \frac{N_A N_B}{1 + \delta_{AB}} \langle \sigma v \rangle_{AB} V^{-1} \tag{B.26b}$$

(b) Neutron–nucleus interaction of type r:

$$r_{r,nF}(\mathbf{x}) = \int_E \Sigma_{r,nF}(E, \mathbf{x}) \Phi(E, \mathbf{x}) \, dE \tag{B.27a}$$

$$R_{r,nF} = \int_V \int_E \Sigma_{r,nF}(E, \mathbf{x}) \Phi(E, \mathbf{x}) \, dE \, d\mathbf{x} \tag{B.27b}$$

(c) Proton-induced spallation interaction

$$r_{pZ}(\mathbf{x}) = n_p(\mathbf{x}) n_Z(\mathbf{x}) \int_{E_p} \tilde{\sigma}_{pZ}(E_p) f_p(E_p, \mathbf{x}) \, dE_p \tag{B.28a}$$

$$R_{pZ} = \int_V n_p(\mathbf{x}) n_Z(\mathbf{x}) \int_{E_p} \tilde{\sigma}_{pZ}(E_p) f_p(E_p, \mathbf{x}) \, dE_p \, d\mathbf{x} \tag{B.28b}$$

Appendix C
Equivalent Fuel Concept

The accounting of fissile fuel is cumbersome because there are several relevant fissile nuclei each possessing its own distinct nuclear properties. If a reactor were supplied with only one type of fissile isotope (index i) and one type of fertile isotope, and if this fertile isotope was the precursor of the fissile isotope i—so that the index $(i-1)$ would be appropriate for the precursor—then the notation for fissile fuel accounting would be particularly simple:

i. the fissile fuel Fi is identical with the isotope i,
ii. the fissile inventory N_{Fi} is given by the number of i-type nuclei $N_{\mathrm{Fi}} = N_i$,
iii. the fissile fuel consumption is described by the number of i-type atoms removed per unit time, $R_{a,\mathrm{Fi}} = R_{a,i}$, and
iv. the fissile fuel generation rate is given by the fertile fuel capture rate $R_{c,\mathrm{Fb}} = R_{c,i-1}$.

As a consequence, the fissile fuel fed into the reactor would be indistinguishable from the fissile fuel discharged from the reactor. Although the concept of a monoisotopic fuel has much appeal and is indirectly widely used, it is clearly unrealistic.

The only naturally occurring fissile isotope is ^{235}U, which is the predominant feed fuel for converter reactors. Except for some special reactors, ^{235}U is presently used in its natural concentration (0.71%) or slightly enriched (2% to 3%) as a reactor fuel. Since ^{235}U has no fertile precursor, it cannot be bred from some other isotope. Nevertheless, the associated ^{238}U can be transmuted into the fissile isotope ^{239}Pu, Eq. (1.3). Thus, in the conventional definition of conversion ratio

$$C_{\text{Fi}} = \frac{\text{fissile nuclei production rate}}{\text{fissile nuclei destruction rate}} \tag{C.1}$$

the fissile nucleus destroyed is ^{235}U from the fuel load or ^{239}Pu bred *in situ* or recycled while the fissile nucleus bred is largely ^{239}Pu.

An additional problem arises since ^{239}Pu is not the only isotopic form in which plutonium exists in a reactor. Some 30% of neutron absorptions in ^{239}Pu yield ^{240}Pu. Because of its large capture cross section, this isotope may be converted to ^{241}Pu, which, in turn, has a high probability for fission as well as for capture leading to ^{242}Pu and higher, nonfissile isotopes. The inclusion of thorium as a fertile fuel further complicates the analysis of fuel cycles.

The fuel balance in a fission reactor is thus characterized by a set of coupled equations for the inventories N_j of each relevant isotope j with the fuel dynamics described by the time variation of the inventory composition vector

$$\mathbf{N} = (N_{232}, N_{233}, \ldots, N_{242}) \tag{C.2}$$

During the time the fuel resides in the reactor, this vector is affected with respect to both its length and direction in its multi-isotope space by the various nuclear reactions. Therefore, there is no unique scalar function $C(t)$ describing the fuel dynamics

$$\mathbf{N}(t) \neq C(t)\mathbf{N}_{\text{load}} \tag{C.3}$$

The one obvious exception is fuel which has been recycled so often that it has reached its equilibrium composition. Excepting this special case, the respective amounts of feed fuel, N_{load}, and discharged fuel, $N_{\text{discharge}}$, can neither be compared to nor subtracted from each other. This has two important consequences:

i. there is no simple scalar fuel trajectory describing the fuel dynamics, and
ii. the definition of simple parameters based on amounts of consumed and produced fuel, such as the conversion ratio C_{Fi}, is complex.

One approach to overcoming this problem is to sum the total number—or, which is more convenient in practical applications, the total mass—of fissile nuclei irrespective of the actual mixture of isotopes present:

$$N_{Fi} = \sum_i N_i, \qquad i = 233, 235, 239, \ldots \tag{C.4}$$

In analogy one defines

$$R_{x,Fi} = \sum_i R_{x,i}, \qquad x = a, c, \text{fi} \ldots ; i = 233, 235, 239, \ldots$$

$$R_{c,Fb} = \sum_i R_{c,i-1}, \qquad i = 233, 239, \ldots$$

$$\text{etc.} \tag{C.5}$$

The shortcoming of this procedure is obvious since it does not incorporate the different nuclear properties of the various isotopes and the pertinent "worth" of each. This suggests that one should not count the number (or mass) of the isotopes but instead measure their respective "worth" or "value" with respect to a particular reactor physics function.

The associated analytical procedure for employing a worth concept consists in defining the principal fissile isotope Fi—with the choice of this reference isotope depending on the fuel cycle in question—and in specifying relative weight factors w_i for each of the other isotopes. That is, w_i isotopes of type i are equivalent to one reference isotope with respect to a given nuclear function.

As an illustration, consider the thermal fission power of a converter reactor. We write for the fission power in a reactor

$$P_{fi,t} = \sum_i U_{r,i} R_{fi,i} \tag{C.6}$$

where $R_{fi,i}$ is the rate at which i-type isotopes undergo thermal and/or fast fission, $U_{r,i}$ is the associated reaction energy, and $i = {}^{235}U$, ${}^{238}U$, ${}^{239}Pu, \ldots {}^{241}Pu$ for a uranium fueled reactor. This may formally be written for an "equivalent" monoisotopic fuel Fi as

$$P_{fi,t} = \tilde{U}_{r,Fi} \tilde{R}_{fi,Fi} \tag{C.7}$$

A comparison between Eqs. (C.6) and (C.7) requires an equivalent reaction rate of

$$\tilde{R}_{fi,Fi} = \frac{1}{\tilde{U}_{r,Fi}} \sum U_{r,i} R_{fi,i}$$

$$= R_{fi,Fi} + \sum_{i \neq Fi} \frac{U_{r,i} R_{fi,i}}{U_{r,Fi}} \tag{C.8}$$

which is the reaction rate for a monoisotopic Fi-fuel yielding the same power as the actual reaction rates in the isotopic mixture. If ^{235}U is taken as the reference fuel isotope, then this equation yields the equivalent fission rate

$$\tilde{R}_{\mathrm{fi,Fi}} = R_{\mathrm{fi},235} + \sum_{i \neq 235} w_{P,i} R_{\mathrm{fi},i} \tag{C.9}$$

Hence, the fission of $w_{P,i} = U_{r,i}/U_{r,235}$ isotopes of type i yields the same contribution to the reactor power as the fission of one ^{235}U isotope. Further, in this particular instance we have according to Table 2.1 the case

$$w_{P,i} \simeq 1, \qquad \text{all } i \tag{C.10}$$

The above example also suggests that there are as many different kinds of sets of equivalent weights w_i as there are nuclear functions associated with the fuel isotopes. It also suggests that there is no single correct set of weights but rather a different set for each application as the next example suggests.

The reactivity–worth equivalent is defined by the requirement that the reactor remains critical if one fuel isotope—the reference fissile isotope, say ^{235}U—is replaced by another fissile isotope or by any adequate mixture of isotopes. The relative reactivity worth of the various fuel isotopes is defined with respect to the principal fertile isotope (^{235}Th or ^{238}U); this is necessary in order to meet the constant-actinide-density constraint condition which requires the removal of a ^{232}Th or ^{238}U atom for each fissile atom added

$$\sum_i N_i = \text{const} \tag{C.11}$$

and the $\rho = 0$ requirement for all fuel composition changes following the initial critical state:

$$\sum_i N_i w_i = 0 \tag{C.12}$$

This constraint is justified from perturbation theory considerations. The w_i are given by

$$\begin{aligned} w_i &= [(\nu_{\mathrm{Fi}} - 1)\sigma_{\mathrm{Fi}} - \sigma_c]_i \\ &= [\nu_{\mathrm{Fi}}\sigma_{\mathrm{Fi}} - \sigma_a]_i \\ &= [\sigma_a(n_{\mathrm{Fi}} - 1)]_i \end{aligned} \tag{C.13}$$

where the nuclear parameters are appropriately space and energy averaged over the neutron distribution in the reaction volume. This indicates again that there is no one universal prescription for calculating the equivalent weight of the relevant isotopes. Since the neutron distribution will change with reactor environment and isotopic mixture, "typical" sets of weights can be given for the various reactor types (LWR, HWR, . . .) and for the various fuel cycle options (once through, Pu recycle, U–Pu recycle . . .).

The reactivity–worth equivalence therefore can be defined by

$$(w_{\mathrm{Fi}} - w_{\mathrm{Fb}})N_{\mathrm{Fi}} = \sum_i (w_i - w_{\mathrm{Fb}})N_i \qquad \text{(C.14)}$$

Apparently isotopic spectral effects are not accounted for in this definition. With

$$w_{\rho,i} = \frac{w_i - w_{\mathrm{Fb}}}{w_{\mathrm{Fi}} - w_{\mathrm{Fb}}}, \qquad w_{\rho,\mathrm{Fi}} = 1, \qquad w_{\rho,\mathrm{Fb}} = 0 \qquad \text{(C.15)}$$

we define the reactivity equivalent weight of the isotope i with respect to the principal fissile element Fi. Therefore replacing one Fi nucleus by w_i i nuclei while removing or adding $|w_i - 1|$ Fb nuclei does not change the criticality state of a reactor because they have the same weight with respect to reactivity ρ.

Therefore, the Fi-equivalent reactivity-weighted inventory N_{Fi} is given by

$$\tilde{N}_{\mathrm{Fi}} = \sum_i w_{\rho,i} N_i \qquad \text{(C.16)}$$

and the equivalent fuel breeding rate is

$$\tilde{R}_{c,\mathrm{Fb}} = \sum_i w_{\rho,i} R_{c,i-1} \qquad \text{(C.17)}$$

and the equivalent fuel reaction rate, for any x-type reaction is

$$\tilde{R}_{x,\mathrm{Fi}} = \sum_i \omega_{\rho,i} R_{x,i} \qquad \text{(C.18)}$$

where $i-1$ and i are the various parent–daughter pairs such as (^{232}Th, ^{233}U), (^{238}U, ^{239}Pu), (^{240}Pu, ^{241}Pu), etc.

Appendix D
Energy Flow Characterization

A variety of expressions may be used to describe energy or power attributes of nuclear reactions and nuclear reactors. Terms such as "power multiplication," "energy gain," "Q value," "break-even," and "ignition" are widely, though often inconsistently, used. Our objective in this Appendix is to develop a more systematic formulation of these related concepts in a compact and analytical form.

The relevant nuclear reactor energy or power characteristics can, in the final analysis, be traced back to reactions of the form

$$A + B \rightarrow \sum_i C_i + U_r \tag{D.1}$$

Here, A, B, and the C_i's are nuclear particles or nucleons and U_r is the nuclear energy released, which may be expressed in terms of the particles masses m_i

$$U_r = \left[(m_A + m_B) - \left(\sum_i m_i\right)\right]c^2 \tag{D.2}$$

where c is the speed of light.

We now introduce two self-evident parameters which may be used to characterize some energy production features of a typical nuclear reaction of interest. An energy multiplication, M_E, is defined by

$$M_E = \frac{U_r}{\langle E_A \rangle + \langle E_B \rangle} \tag{D.3}$$

and the energy gain, G_E, is given by

$$G_E = \frac{U_r - [\langle E_A \rangle + \langle E_B \rangle]}{\langle E_A \rangle + \langle E_B \rangle} \tag{D.4}$$

Here the symbol $\langle \ \rangle$ is used to indicate the appropriate average kinetic energies of the particles A and B.

Evidently, M_E and G_E are related by

$$G_E = M_E - 1 \tag{D.5}$$

Typical numerical values for energy multiplication for the three kinds of nuclear reactions of interest here are displayed in Table D.1.

We alert the reader here to the feature that both M_E and G_E—while possessing a symbolism convenient to be associated with their physical meaning—represent simply ratios of energy. Such dimensionless energy ratios can be constructed for innumerable cases of interest and can be the source of much confusion. We attempt an elaboration of these ideas beginning with its extension to power.

Power multiplication and power gain expressions are readily accomplished by noting that the instantaneous power P_i associated with a reaction rate of type R_i with an associated average energy release U_i per reaction is given by

$$P_i = R_i U_i, \qquad i = fi, fu, sp, \ldots \tag{D.6}$$

TABLE D.1
Ranges of Energy Multiplication in Fusion, Fission, and Spallation Processes

Reaction	M_E
Fission (thermal and fast reactors)	10^3 to 10^9
Fusion (D–T, D–D)	10^1 to 10^3
Spallation	$<10^0$

With R_i in units of s^{-1} and U_i in units of MeV, the power P_i possesses units of MeV/s, which can be converted to any other convenient units of power. Eq. (D.6) may be directly associated with the power output within a reaction domain { } and represented by

$$\{\text{reaction domain}\} \rightarrow P_{\text{out},i} = R_i U_i \tag{D.7}$$

Many reaction domains, such as fusion chambers and spallation targets, require a power input in order to sustain the reaction rate R_i. We extend Eq. (D.7) therefore to read

$$P_{\text{in},i} \rightarrow \{\text{reaction domain}\} \rightarrow P_{\text{out},i} \tag{D.8}$$

The instantaneous power multiplication M_p and power gain G_p are by analogy to Eq. (D.3) and Eq. (D.4) given by

$$M_p = \frac{\sum_i P_{\text{out},i}}{\sum_i P_{\text{in},i}} = \frac{\sum_i R_i U_i}{\sum_i P_{\text{in},i}} \tag{D.9}$$

and

$$G_p = \frac{\sum_i (P_{\text{out},i} - P_{\text{in},i})}{\sum_i P_{\text{in},i}} = M_p - 1 \tag{D.10}$$

Often, a more detailed power balance than suggested by Eq. (D.8) may need to be undertaken. In some cases, the fraction of the nuclear power retained *in situ* may be important in sustaining subsequent reactions; a case in point is the heating of a D–T plasma with 3.5-MeV alpha reaction product. Additionally, some of the energy in the reaction domain may be lost to unrecoverable forms of energy. Hence, a more appropriate characterization needs to be introduced and is depicted in Figure D.1.

In Figure D.1, we also refer to different forms of power. At the fundamental level we have the nuclear reaction power U_r and the fraction of the power entering the reaction domain and directly transformed into kinetic energy of reacting particles, $P_{\text{in},r}$. The associated power balance is

$$P_{\text{in},r} + R_r U_r = P_{\text{out},r} \tag{D.11}$$

Here, the subscript r refers to the particular nuclear reaction. The component $P_{\text{ret},r}$ in Figure D.1 is that part of $P_{\text{out},r}$ which is retained in the

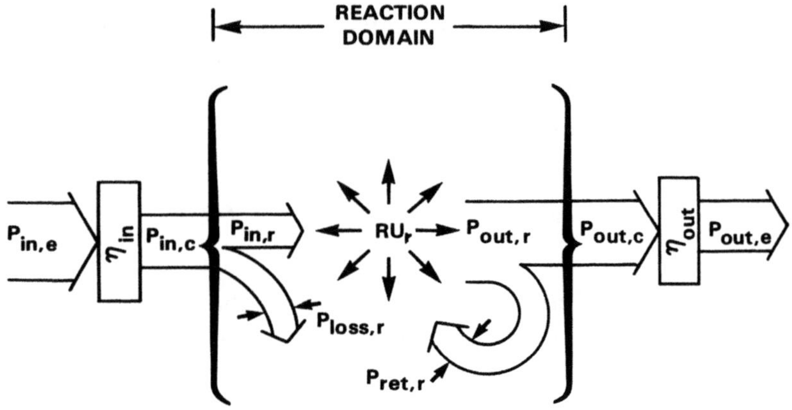

FIGURE D.1. Depiction of power flow in a nuclear reaction domain. The widths of the arrows are schematic only and are not meant to represent magnitudes of power.

reaction domain and the term $P_{\text{loss},r}$ is the part of the input power which does not contribute to the reaction rate R_r.

At the next level of emphasis we consider core power. In Figure D.1, $P_{\text{in},c}$ is the power which is absorbed in the reaction core domain while $P_{\text{out},c}$ is the core power released from the reaction domain and available for transformation into electrical form. The associated electrical forms of power are related to the other forms of power by the following relations:

$$\begin{aligned} \eta_{\text{in}} P_{\text{in},e} &= P_{\text{in},c} \\ &= P_{\text{in},r} + P_{\text{loss},r} \end{aligned} \quad \text{(D.12)}$$

$$\begin{aligned} P_{\text{out},e} &= \eta_{\text{out}} P_{\text{out},c} \\ &= \eta_{\text{out}}(P_{\text{out},r} - P_{\text{ret},r}) \end{aligned} \quad \text{(D.13)}$$

Here, the electrical-to-core and core-to-electrical conversion efficiencies are η_{in} and η_{out}, respectively.

We now recognize three dimensionless power multiplications. At the nuclear reaction level, we write

$$M_r = \frac{P_{\text{out},r}}{P_{\text{in},r}} \quad \text{(D.14)}$$

which essentially reproduces the basic nuclear reaction energy multiplication, Eq. (D.3). Further, a core power multiplication is defined by

$$M_c = \frac{P_{\text{out},c}}{P_{\text{in},c}} \tag{D.15}$$

and finally, an electrical power multiplication is given by

$$M_e = \frac{P_{\text{out},e}}{P_{\text{in},e}} \tag{D.16}$$

These three power multiplications are clearly not independent because of the several power relationships which exist. For example, by extension of Eq. (D.15)

$$M_c = \frac{P_{\text{out},r} - P_{\text{ret},r}}{P_{\text{in},r} + P_{\text{loss},r}} \tag{D.17}$$

and if $P_{\text{loss},r} \ll P_{\text{in},r}$ we get

$$M_c \simeq M_r - \frac{P_{\text{ret},r}}{P_{\text{in},r}} \tag{D.18}$$

Then, M_e can be related to M_c by extension of Eq. (D.16)

$$M_e = \frac{\eta_{\text{out}} P_{\text{out},c}}{P_{\text{in},c}/\eta_{\text{in}}}$$

$$= \eta_{\text{in}}\eta_{\text{out}} M_c \tag{D.19}$$

With $\eta_{\text{in}}\eta_{\text{out}} < 1$, the general ordering is evidently

$$M_e < M_c < M_r \tag{D.20}$$

and it generally becomes very important to distinguish between these power multiplications.

We consider another relevant extension. As encountered repeatedly, many nonfission reactions produce neutrons used to breed fissile nuclei.

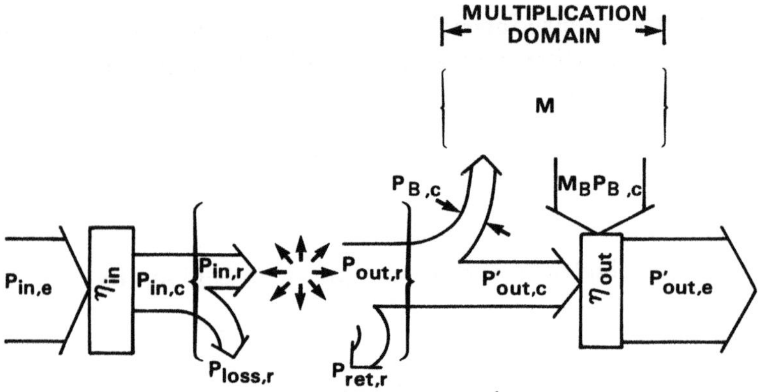

FIGURE D.2. Schematic illustration showing power multiplier effect associated with the use of nonfission neutrons for fissile fuel breeding.

These nuclei eventually fission in the same or in a separate device yielding another energy/power multiplier effect. We illustrate Figure D.2 how the fraction of core power $P_{m,c}$ is multiplied by M_B for subsequent conversion into electrical form.

Based on Figure D.2, the electrical power multiplication is now

$$M'_e = \frac{P'_{\text{out},e}}{P_{\text{in},e}} \tag{D.21}$$

The role of the fission multiplication effect can be identified by converting these electrical powers into their associated nuclear components which are identical for both Figure D.1 and Figure D.2. Some algebraic manipulation leads to the following relation:

$$P'_{\text{out},e} = \eta_{\text{out}}[P_{\text{out},r} - P_{\text{ret},r} + P_{B,c}(M_B - 1)] \tag{D.22}$$

and, from Eq. (D.12),

$$P_{\text{in},e} = \frac{1}{\eta_{\text{in}}}[P_{\text{in},r} + P_{\text{loss},r}] \tag{D.23}$$

From this it follows that

$$M'_e = \eta_{\text{in}}\eta_{\text{out}}\left[M_c + (M_B - 1)\left(\frac{P_{B,c}}{P_{\text{in},r} + P_{\text{loss},r}}\right)\right] \tag{D.24}$$

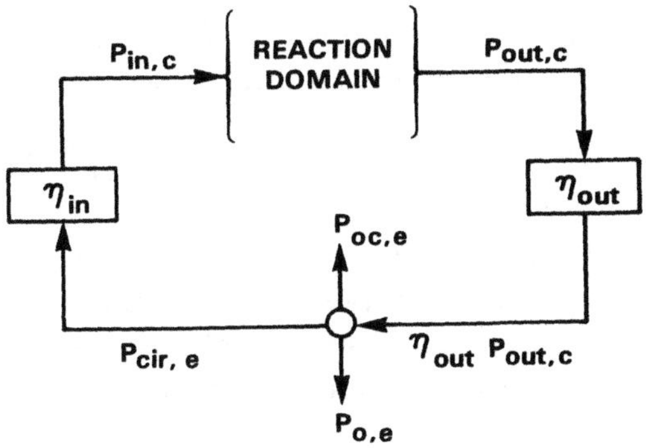

FIGURE D.3. General schematic illustrating gross power flow in a nuclear power station.

Clearly, for $M_B > 1$, which is readily feasible, we have therefore

$$M_e' > M_e \tag{D.25}$$

The preceding decomposition of the power flows and the various power ratios and expressions now allows for a compact representation of the nuclear power as frequently found in the literature is schematically depicted in Figure D.3. We write the station net electrical output $P_{o,e}$ as

$$P_{o,e} = \eta_{out} P_{out,c} - P_{ci,e} - P_{oc,e} \tag{D.26}$$

where $P_{out,c}$ is the reactor core output power, $P_{ci,e}$ is the electrical circulating power required to maintain the reaction, and $P_{oc,e}$ in the electrical requirement for station operation and control.

It is common in the literature to define "Q-values" for reactions, fusion plasmas, and fusion reactors. Usually, the Q-value of a reaction corresponds to what we have here defined as the reaction energy

$$Q_r = U_r \tag{D.27}$$

The Q-value of the plasma is generally defined as

$$Q_p = \frac{\text{fusion power yield}}{\text{external power deposited}}$$

$$= \frac{P_{out,r}}{P_{in,c}} = M_c + \frac{P_{ret,r}}{P_{in,c}} \tag{D.28}$$

while the Q-value of the fusion reactor is usually defined in the same way as the corresponding electric power multiplication parameter:

$$Q_e = M_e \tag{D.29}$$

This emphasizes the importance of the appropriate dimensionless energy/power expressions as parameters of merit of a system.

As a final point, we refer to the term "ignition." This terms applies to a reaction domain for which

$$P_{\text{in},c} = 0 \qquad (\text{i.e., } P_{\text{ci},e} = 0) \tag{D.29}$$

and hence implies that $M_c = \infty$ and also $M_e = \infty$.

We conclude then that a fission reactor is therefore ignited with an infinite "Q value," a spallation accelerator is a device with a small "Q value" and hence subignited, while a fusion device can in principle cover the entire range of this "Q value" parameter.

Index

GPSR Compliance

The European Union's (EU) General Product Safety Regulation (GPSR) is a set of rules that requires consumer products to be safe and our obligations to ensure this.

If you have any concerns about our products, you can contact us on ProductSafety@springernature.com

In case Publisher is established outside the EU, the EU authorized representative is:

Springer Nature Customer Service Center GmbH
Europaplatz 3
69115 Heidelberg, Germany

Batch number: 10615115

Printed by Printforce, the Netherlands